Commission of the European Communities

Patent Information and Documentation in Western Europe

An Inventory of Services available to the Public

Third Edition

Edited by
Brenda M. Rimmer

K · G · Saur
München · New York · London · Paris 1988

EUR 6614 (1988 Edition)

The Commission of the European Communities
Directorate-General Telecommunications, Information Industries and Innovation

Jean Monnet Building Kirchberg
Luxembourg
Tel. (0352) 43011

Title of the French edition: Information et documentation en matière de brevets en Europe
  occidentale, Inventaire des services offerts au public.
Title of the German edition: Patentinformation und Patentdokumentation in Westeuropa.
  Ein Bestandsverzeichnis von öffentlich zugänglichen Diensten.

LEGAL NOTICE

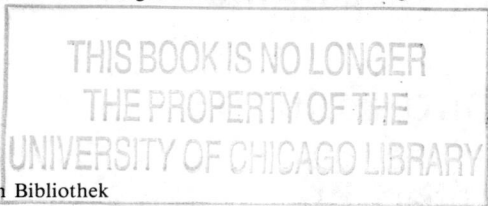

CIP-Kurztitelaufnahme der Deutschen Bibliothek

**Patent information and documentation in
Western Europe** : an inventory of services
available to the public / Comm. of the
Europ. Communities. [The Comm. of the
Europ. Communities, Directorate General
"Information Market and Innovation"].
Ed. by: Brenda M. Rimmer − 3., rev. and enl. ed.
− München ; London ; New York ; Paris :
Saur, 1988.
  Dt. Ausg. u.d.T.: Patentinformation und
  Patentdokumentation in Westeuropa. −
  Franz. Ausg. u.d.T.: Information et
  documentation en matière de brevets en
  Europe occidentale

  ISBN 3-598-10744-7

NE: Rimmer, Brenda M. [Hrsg.]; Europäische
Gemeinschaften / Kommission

© ECSC-EEC-EAEC, Brussels-Luxembourg, 1988
Publisher: K. G. Saur Verlag, München
(A member of the international Butterworth Group, London)
Printed by Weihert Druck GmbH, Darmstadt
Bound by Kränkl, Heppenheim
Federal Republic of Germany
ISBN 3-598-10744-7

CONTENTS

INTRODUCTION TO THE THIRD EDITION

Since the second edition of this publication in 1981 there have been few major changes in legislation but an increasing awareness of the importance of patent documentation as a source of technological information has encouraged many authorities to improve the quality of their published literature and extend the facilities available to the public. Supported by computer based searching aids there are now few patent offices which do not offer professional assistance to inventors, repayment being usually on the basis of time taken, with the option of a fixed maximum cost. Most patent offices now have their own database and also access to the growing number of commercially available databases.

The contributing countries and authorities are the same as in the second edition but they are now arranged in order of country code instead of by name and thus appear in the same sequence in all language editions. The lists of holdings of the various libraries are also in order of country code. With a few exceptions, the front page of only the most significant patent specification for each country has been reproduced. Included for reference as before at the end are brief details of the publications of Japan, the Soviet Union and the United States of America.

The information is arranged under the same paragraph headings for each country, viz.,

I   General information
II   Patent documents
III   Official gazette
IV   Sources of supply and prices
V   Register of legal status
VI   Public services
VII   Provincial libraries

No detailed lists of information in the registers of legal status are given as the data available was much the same in all countries, i.e. bibliographic details, stages of progress, payments made, legal proceedings, etc. The lists of library holdings have been abbreviated as it was felt that the dates of earliest specifications (specns) and official gazettes (gazette) would provide a suffiently reliable indication of the scope of the relevant library. It can be assumed, for example, that a library holding DE specifications from 1877 would also hold current *Offenlegungsschriften*. Full titles of the specifications and official gazettes are included for reference in the list of holdings of the Science Reference and Information Service (SRIS) of the British Library (formerly the Science Reference Library).

Throughout EPC and PCT denote European Patent Convention and Patent Cooperation Treaty respectively and the currently preferred European term "Telecopier" has been used for "Fax", "Facsimile" and "Telefax".

Chemical Abstracts Service (CAS), Derwent Publications and INPADOC each have a separate chapter and there is a summary of commercial databases devoted to patents as well as a list of journals which include patent abstracts in their coverage, some of which are also accessible online.

A general index has been included.

The Editor

STANDARDS FOR PATENT LITERATURE

When the Committee for International Cooperation in Information Retrieval among Patent Offices (ICIREPAT) ceased to exist in 1979 its functions were transferred to the World Intellectual Property Organization (WIPO) Permanent Committee on Patent Information (PCPI).

The *ICIREPAT manual* has been succeeded by the *Patent Information and Documentation Handbook* first published by WIPO in 1981. It is in loose leaf form and regularly updated, the fifth issue in 1985 needing the introduction of a fourth binder. The text is bilingual English/French.

An important part of the publication is devoted to the standards and guidelines facilitating the identification and understanding of patent documentation. These have been adopted by most countries in the world and, by permission of WIPO, parts of two standards are reproduced in following pages for convenient reference.

These are:

ST.3 A two letter "country code" identifying states and other entities which issue patent documents.

ST.9 A numerical code relating to the bibliographic data on patent documents and usually referred to as the "INID code" (Internationally agreed Numbers for the Identification of Data)

Another WIPO standard used in the identification of patent documents is ST.16 which defines the status of the letter code placed after the serial number.

The suffixes A, B and C are generally used to identify the level of publication of the patent document (U and Y refer to utility models). For example GB 2 123 456A defines an unexamined GB specification published eighteen months after its filing or priority date while the same number with a suffix B refers to an examined and granted patent. In the Federal Republic of Germany where there were previously three levels of routine publication, B has now been discontinued, leaving A as the first level and C as the second.

An additional numerical suffix is used on European patent and PCT publications to indicate the availability of the search report, i.e. A1 with search report, A2 without search report and A3 separate publication of the search report.

Other reference books published by WIPO, both loose leaf for convenient updating, are:

World Directory of Sources of Patent Information 1985
This consistes mainly of data sheets for most of the WIPO member countries with tabulated lists of the documents available at each of the information sources covered, principally patent offices, together with various indexes.

Directory of National and Regional Industrial Property Offices with addresses
and telephone/telex numbers.

List of Countries, and of Other Entities
Issuing or Registering Industrial Property Titles
(in the order corresponding to the current UN and WIPO practice)

Afghanistan	AF	Comoros	KM
Albania	AL	Congo	CG
Algeria	DZ	Costa Rica	CR
Angola	AO	Côte d'Ivoire	CI
Anguilla	AI	Cuba	CU
Antigua and Barbuda	AG	Cyprus	CY
Argentina	AR	Czechoslovakia	CS
Australia	AU		
Austria	AT	Democratic Kampuchea	KH
		Democratic People's Republic	
Bahamas	BS	of Korea	KP
Bahrain	BH	Democratic Yemen	YD
Bangladesh	BD	Denmark	DK
Barbados	BB	Djibouti	DJ
Belgium	BE	Dominica	DM
Belize	BZ	Dominican Republic	DO
Benin	BJ		
Bermuda	BM	Ecuador	EC
Bhutan	BT	Egypt	EG
Bolivia	BO	El Salvador	SV
Botswana	BW	Equatorial Guinea	GQ
Brazil	BR	Ethiopia	ET
British Virgin Islands	VG		
Brunei Darussalam	BN	Falkland Islands (Malvinas)	FK
Bulgaria	BG	Fiji	FJ
Burkina Faso	BF	Finland	FI
Burma	BU	France	FR
Burundi	BI		
		Gabon	GA
Cameroon	CM	Gambia	GM
Canada	CA	German Democratic Republic	DD
Cape Verde	CV	Germany, Federal Republic of	DE
Cayman Islands	KY	Ghana	GH
Central African Republic	CF	Gibraltar	GI
Chad	TD	Greece	GR
Chile	CL	Grenada	GD
China	CN	Guatemala	GT
Colombia	CO	Guinea	GN

Guinea-Bissau	GW	Nauru	NR
Guyana	GY	Nepal	NP
		Netherlands	NL
Haiti	HT	Netherlands Antilles	AN
Holy See	VA	New Zealand	NZ
Honduras	HN	Nicaragua	NI
Hong Kong	HK	Niger	NE
Hungary	HU	Nigeria	NG
		Norway	NO
Iceland	IS		
India	IN	Oman	OM
Indonesia	ID		
Iran (Islamic Republic of)	IR	Pakistan	PK
Iraq	IQ	Panama	PA
Ireland	IE	Papua New Guinea	PG
Israel	IL	Paraguay	PY
Italy	IT	Peru	PE
		Philippines	PH
Jamaica	JM	Poland	PL
Japan	JP	Portugal	PT
Jordan	JO		
		Qatar	QA
Kenya	KE		
Kiribati	KI	Republic of Korea	KR
Kuwait	KW	Romania	RO
		Rwanda	RW
Laos	LA		
Lebanon	LB	Saint Christopher and Nevis	KN
Lesotho	LS	Saint Helena	SH
Liberia	LR	Saint Lucia	LC
Libya	LY	Saint Vincent and the	
Liechtenstein	LI	Grenadines	VC
Luxembourg	LU	Samoa	WS
		San Marino	SM
Madagascar	MG	Sao Tome and Principe	ST
Malawi	MW	Saudi Arabia	SA
Malaysia	MY	Senegal	SN
Maldives	MV	Seychelles	SC
Mali	ML	Sierra Leone	SL
Malta	MT	Singapore	SG
Mauritania	MR	Solomon Islands	SB
Mauritius	MU	Somalia	SO
Mexico	MX	South Africa	ZA
Monaco	MC	Soviet Union	SU
Mongolia	MN	Spain	ES
Montserrat	MS	Sri Lanka	LK
Morocco	MA	Sudan	SD
Mozambique	MZ	Suriname	SR

Swaziland	SZ	United Kingdom	GB
Sweden	SE	United Republic of Tanzania	TZ
Switzerland	CH	United States of America	US
Syria	SY	Uruguay	UY
Taiwan, Province of China	TW	Vanuatu	VU
Thailand	TH	Venezuela	VE
Togo	TG	Viet Nam	VN
Tonga	TO		
Trinidad and Tobago	TT	Yemen	YE
Tunisia	TN	Yugoslavia	YU
Turkey	TR		
Tuvalu	TV	Zaire	ZR
		Zambia	ZM
Uganda	UG	Zimbabwe	ZW
United Arab Emirates	AE		

International organizations
issuing or registering industrial property titles

African Intellectual Property Organization (OAPI)	OA
African Regional Industrial Property Organization (ARIPO)	AP
Benelux Trademark Office and Benelux Designs Office	BX
European Patent Office (EPO)	EP
World Intellectual Property Organization (WIPO)	WO

INID code relating to bibliographic data on patent documents

(10) *Document identification*
  (11) Number of the document
  (12) Plain language designation of the kind of document
  (13) Kind of document code according to WIPO Standard ST.16
  (19) WIPO Standard ST.3 code, or other identification, of the office publishing the document
Notes: (i) Minimum data element for patent documents only
         (ii) with the proviso that when data coded (11) and (13), or (19), (11) and (13), are used together and on a single line, category (10) can be used, if so desired

(20) *Domestic filing data*
  (21) Number(s) assigned to the application(s), e.g. "Numéro d'enregistrement national", "Aktenzeichen"
  (22) Date(s) of filing application(s)
  (23) Other date(s) of filing, including exhibition filing date and date of filing complete specification following provisional specification
  (24) Date from which industrial property rights may have effect
  (25) Language in which the published application was originally filed
  (26) Language in which the application is published.

(30) *Priority Data*
  (31) Number(s) assigned to priority application(s)
  (32) Date(s) of filing of priority application(s)
  (33) WIPO Standard ST.3 Code identifying the national patent office allotting the priority application number or the organization allotting the international or regional priority application number

Notes: (i) With the proviso that when data coded (31), (32) and (33) are used together and on a single line, category (30) can be used, if so desired.
         (ii) The presentation of priority application numbers should be as recommended in WIPO Standards ST.10/C and in ST.34.

(40) *Date(s) of making available to the public*
  (41) Date of making available to the public by viewing, or copying on request, an *unexamined* document, on which no grant has taken place on or before the said date
  (42) Date of making available to the public by viewing, or copying on request, an *examined* document, on which no grant has taken place on or before the said date
  (43) Date of publication by printing or similar process of an *unexamined* document, on which no grant has taken place on or before the said date
  (44) Date of publication by printing or similar process of an *examined* document, on which no grant has taken place on or before the said date
  (45) Date of publication by printing or similar process of a document on which grant has taken place on or before the said date

13

(46) Date of publication by printing or similar process of the claim(s) only of a document

(47) Date of making available to the public by viewing, or copying on request, a document on which grant has taken place on or before the said date

Note: Minimum data element for patent documents only, the minimum data requirement being met by indicating the date of making available to the public the document concerned

(50) *Technical information*
(51) International Patent Classification
(52) Domestic or national Classification
(53) Universal Decimal Classification
(54) Title of the invention
(55) Keywords
(56) List of prior art documents, if separate from descriptive text
(57) Abstract or claim
(58) Field of search

(60) *References to other legally related domestic patent documents including unpublished applications therefor*
(61) Number and, if possible, filing date of the earlier application, or number of the earlier publication, or number of earlier granted patent, inventor's certificate, utility model or the like to which the present document is an addition
(62) Number and, if possible, filing date of the earlier application from which the present document has been divided out
(63) Number and filing date of the earlier application of which the present document is a continuation
(64) Number of the earlier publication which is "reissued"
(65) Number of a previously published patent document concerning the same application

Notes: (i) Priority data should be coded in category (30)
(ii) Code (65) is intended primarily for use by countries in which the national laws require that re-publication occurs at various procedural stages under different publication numbers and these numbers differ from the basic application numbers

(70) *Identification of parties concerned with the document*
(71) Name(s) of applicant(s)
(72) Name(s) of inventor(s) if known to be such
(73) Name(s) of grantee(s)
(74) Name(s) of attorney(s) or agent(s)
(75) Name(s) of inventor(s) who is (are) also applicant(s)
(76) Name(s) of inventor(s) who is (are) also applicant(s) and grantee(s)

Notes (i) For documents on which grant has taken place on or before the date of making available to the public, and gazette entries relating thereto, the minimum data requirement is met by indicating the grantee, and for other documents by indicating the applicant.

(ii)   (75) and (76) are intended primarily for use by countries in which the national laws required that the inventor and applicant are normally the same. In other cases (71) or (72) or (71), (72) and (73) should generally be used.

(80)   *Identification of data related to International Conventions other than the Paris Convention*

    (81)   Designated State(s) according to the PCT

    (83)   Information concerning the deposit of microorganisms, e.g. under the Budapest Treaty

    (84)   Designated contracting states under regional patent conventions

    (85)   Date of fulfillment of the requirements of articles 22 and/or 39 of the PCT for introducing the national procedure according to the PCT

    (86)   Filing data of the regional or PCT application, i.e. application filing date, application number, and, optionally, the language in which the published application was originally filed

    (87)   Publication data of the regional or PCT application, i.e. publication date, publication number, and, optionally, the language in which the application is published

    (88)   Date of deferred publication of the search report

    (89)   Document number and country of origin of the original document according, to the CMEA Agreement on Mutual Recognition of Inventors' Certificates and other Documents of Protection for Inventions

Notes: (i)   The codes (86) and (87) are intended to be used:

        − on *national* documents when identifying one or more of the relevant filing data or publication data of a *regional* or *PCT* application, or

        − on *regional* documents when identifying one or more of the relevant filing data or publication data of *another regional* or *PCT* application.

    (ii)   all data in code (86) should be presented together and preferably on a single line.

    (iii)   all data in code (87) should also be presented together and preferably on a single line.

PATENT OFFICE

ÖSTERREICHISCHES PATENTAMT
Kohlmarkt 8 – 10
A – 1014 Vienna
Tel: (0222) 63 36 36
Telex. 136847 OEPA A

I General information

The Austrian Patent Office is responsible to the Minister for Commerce, Trade and Industry (*Bundesminister für Handel, Gewerbe und Industrie)* for the protection of industrial property rights. It is an examining patent office and is an International Searching Authority under the PCT for a number of countries. It provides reports on state of the art, patentability and related problems. This service is assured by nineteen technical, an appeal and two legal departments.

Staff (total):	272
Patent examiners:	127
Legal experts:	17
Library and documentation:	11

II Patent documents

Printed specifications of granted patents
1899 – 1944   No. 1 – 160 924
1949 –   No. 162 001 –

III Official gazette

Österreichisches Patentblatt   1899 –   monthly (on the 15th)

Part I
Laws, regulations and notices relating to industrial property rights, patent office notices, decisions of the Patent Office Court and of the Supreme Patent and Trade Mark Chamber (*Oberster Patent- und Markensenat*) and other general information. The Patent Office report and annual statistics are in the March issue.

ÖSTERREICHISCHES   ⑤ Int.Cl⁴: C07C 045/40
PATENTAMT

⑲ AT PATENTSCHRIFT   ⑪ Nr. 380 008

⑦ Patentinhaber: CHEMIE LINZ AKTIENGESELLSCHAFT
LINZ, OBERÖSTERREICH

⑤ Gegenstand: VERFAHREN ZUR HERSTELLUNG VON MONO- ODER
BISCARBONYLVERBINDUNGEN

⑥ Zusatz zu Patent Nr.
⑥ Ausscheidung aus:
㉒㉑ Angemeldet: 1983 12 23,   4500/83
㉓ Ausstellungspriorität:

㉝㉜㉛ Unionspriorität:

㊷ Beginn der Patentdauer: 1985 08 15
Längste mögliche Dauer:
㊺ Ausgegeben: 1986 03 25
⑦ Erfinder: SAJTOS ALEXANDER DIPL.ING.
LINZ, OBERÖSTERREICH
WECHSBERG MANFRED DIPL.ING. DR.
LINZ, OBERÖSTERREICH
ROITHNER ERICH
LINZ, OBERÖSTERREICH
POLLHAMMER STEFAN
LINZ, OBERÖSTERREICH
MAHRINGER ANDREAS
LINZ, OBERÖSTERREICH

⑥ Abhängigkeit:

㊺ Druckschriften, die zur Abgrenzung vom Stand der Technik in Betracht gezogen wurden:

DE-OS 2514001   GB-PS 709450   US-PS 3705922   US-PS 3637721
US-PS 3145232

AT 380 008

18

Part II

Application number and bibliographic details of accepted applications laid open to public inspection and also of granted patents arranged by class.

There is an annual index to each part published at the beginning of the following year.

IV Source of supply and prices (from 1. 1. 84)

Specifications, photocopies and the official gazette are obtainable from the Patent Office.

	Price öS
Specifications	40.00
Photocopies of foreign patents,	
per page, delivery in two days	6.60
rapid service	9.70
Official gazette	
Part 1 single copy	45.00
annual subscription	500.00
Part II single copy	120.00
annual subscription	1280.00
Parts I and II	
annual subscription	1580.00

The annual subscription includes the indexes which are not available separately.

V Register of legal status

Bibliographic data relating to Austrian patent specifications, from the first publication in 1899, are recorded in the *Patentregister* together with their current legal status. The *Patentregister*, which was the first computer produced book in the public sector in Austria, is arranged in numerical order and is freely accessible to the general public.

VI Public services

The literature of the Austrian Patent Office is available in the public reading room.

Open: Monday to Friday 8.00 to 14.00

The library holds about 29 million patent documents from 28 countries and 4 international organizations, 41 official gazettes, 466 current technical and legal journals, about 30,000 volumes of technical and legal books and about 90,000 early patents (*Privilegien*).

The patent specifications are arranged numerically for ready retrieval and the IPC with its associated keyword index enables subject searches to be made and the current state of the technology to be assessed.

Patent specifications and official gazettes held in the library of the Austrian Patent Office

AP	ESARIPO	specns	1985 –
		gazette	1985 –
AT	Austria	specns	1899 –
		gazette	1899 –
AU	Australia	specns	1926 –
		gazette	1905 –
BE	Belgium	specns	1950 –
		gazette	1866 –
BG	Bulgaria	specns	1962 –
		gazette	1964 –
BR	Brazil	gazette	1972 –
CA	Canada	specns	1950 –
		gazette	1873 –
CH	Switzerland	specns	1888 –
		gazette	1889 –
CN	China	specns	1985 –
		gazette	1985 –
CO	Columbia	gazette	1958 –
CS	Czechoslovakia	specns	1920 –
		gazette	1919 –
CU	Cuba	gazette	1961 –
DD	German Democratic Republic	specns	1956 –
		gazette	1960 –
DE	Germany, Federal Republic of	specns	1877 –
		gazette	1877 –
DK	Denmark	specns	1894 –
		gazette	1900 –
DZ	Algeria	gazette	1978 –
EP	European Patent Office	specns	1978 –
		gazette	1978 –
ES	Spain	specns	1984 –
		gazette	1887 –
FI	Finland	specns	1963 –
		gazette	1963 –
FR	France	specns	1902 –
		gazette	1884 –
GB	United Kingdom	specns	1617 –
		gazette	1854 –

GR	Greece	gazette	1966 –
HU	Hungary	specns	1896 –
		gazette	1896 –
IE	Ireland	gazette	1960 –
IL	Israel	gazette	1963 –
IT	Italy	specns	1926 –
		gazette	1902 –
JP	Japan	specns	1974 –
KR	Korea, Republic of	gazette	1979 –
MX	Mexico	gazette	1903 –
NL	Netherlands	specns	1913 –
		gazette	1912 –
NO	Norway	specns	1892 –
		gazette	1911 –
NZ	New Zealand	gazette	1947 –
OA	OAPI	specns	1966 –
		gazette	1966 –
PL	Poland	specns	1924 –
		gazette	1924 –
RO	Romania	specns	1957 –
		gazette	1968 –
SE	Sweden	specns	1885 –
		gazette	1946 –
SU	Soviet Union	specns	1924 –
		gazette	1956 –
TR	Turkey	gazette	1931 –
TW	Taiwan	gazette	1974 –
US	United States of America	specns	1871 –
		gazette	1872 –
VN	Vietnam, Socialist Republic of	specns	1985 –
		gazette	1985 –
WO	WIPO-PCT	specns	1978 –
		gazette	1978 –
YU	Yugoslavia	specns	1922 –
		gazette	1921 –

Also available
- Abstracts of German utility models
  (*Auszüge aus den Gebrauchsmustern*)   1970 –
- Patent Abstracts of Japan   1977 –
- Derwent abstracts in English for Japan (chemical) and the Soviet Union
- Online access to WPI and WPIL (Derwent); CAS (Chemical Abstracts Service) from Télésystèmes; INPADOC (PFS)

Special services

While searches may be made in person in the Austrian patent office library experts are available to supply information in writing on the following aspects: –

Reports on
- State of the art searches with respect to a specific technical problem, fee öS 2000.
- The patentability of an invention. The fee is öS 2000 if the current state of the art is disclosed by the inventor or öS 3000 if it is required to be researched by the patent office.

Bibliographical information on
- Applications filed during a certain period of time in a specific subject field, by a particular applicant or by a particular inventor.
- Similar information on applications laid open to public inspection.
- Details from the public patent register on Austrian patents and European patents effective in Austria.

For written information services based on automated systems there is a basic charge of öS 50 and a line charge of öS 30 per 100 lines of print. For written information not based on automated systems on up to three patent applications or granted patents the charge is öS 30.

Publicly accessible databanks can provide the following details: –
- application date, title, name of applicant (and agent), filing number
- IPC
- priority details
- stage of progress

Bibliographic details of Austrian applications
- in a specific field of technology
- on a particular applicant or inventor
- in a stated period of time

Bibliographic details of applications laid open to public inspection in Austria and European and International applications designating Austria and of European patents effective in Austria
- in a specific subject field
- on a particular applicant or inventor

Legal experts on the staff of the Patent Office provide free information on all aspects of industrial property rights on Monday, Tuesday, Thursday and Friday 10.00 – 12.00.

VII Provincial libraries

Both parts of the *Patentblatt* are available to the public in the libraries of the Chambers of Trade and Industry (Landeskammern der gewerblichen Wirtschaft) in Vienna, Linz, Salzburg, Innsbruck, Dornbirn and Graz.

Austrian patent specifications are also available in Linz, Innsbruck and Dornbirn and in the technical universities of Vienna and Graz.

BELGIUM

BE

PATENT OFFICE

OFFICE DE LA PROPRIÉTÉ INDUSTRIELLE (OPRI)
(Ministère des Affaires économiques)
24 – 26, rue J. A. De Mot
B – 1040 Brussels
Tel: 233 61 11
Telex: 20627 COMHAN B

I   General information

The *Office de la Propriété Industrielle (OPRI)* is a national service under the authority of the Ministre des Affaires economiques (Minister for Economic Affairs) and is responsible for the granting of patents in Belgium. Examination is for formalities only. In accordance with the new patent law of 28 March 1984 (in force from 1 January 1987) a search report established by the European Patent Office may be requested. In this case the patent has a maximum duration of 20 years from its application date, but otherwise, the maximum duration is 6 years.

Staff (total including library): 65

II   Patent documents

The Belgian patent files are laid open to public inspection on the day the patent is granted. This is as soon as possible after a delay of 18 months from the application date or the priority date. The applicant may request grant as soon as the formalities have been completed.

The patent specifications from 1 January 1987 comprise, except for rare instances of publication before grant, the research report, the final version of the description, claims, drawings and the certificate of grant.

The specifications are not printed but copies may be obtained in the following forms: –
– photocopies (may be certified true copies if required)
– microfiches from no. 895 529 (1983)
– aperture cards, 796 001 – 895 528 (1973 – 83) when available

23

Publication is in the language used by the applicant conforming with the Belgian legislation concerning the use of official languages. OPRI does not provide translation of the text and claims. In practice most of the specifications are in French with, currently, about 20 % in Dutch and a few in German.

III Official gazette

Recueil des Brevets d'Invention   1854 –   monthly
Contents
- abstract and bibliographic details
- heading data for European patents effective in Belgium
- name index to patentees
- numerical index
- stages of progress

An additional annual publication containing
- classified list of patent applications made during the year
- name index to patentees
- patents assigned or licenced
- patents lapsed due to non payment of fees
- lapsed patents restored
- numerical list with reference to related entry in the official gazette

IV Sources of supply and prices

Publications and copies of specifications may be obtained from the Patent Office as well as photocopies of foreign patent specifications and other publications. Documents requested must be clearly defined by country and number.

	Prices from 1. 1. 87 BF
Recueil des brevets d'invention	
Annual subscription	2000
Postage: to Belgium	750
to Europe	2000
outside Europe	10000
Copies	
photocopies per page	15
photocopies per page, urgent despatch	30
microfiches, Belgian patents, per microfiche	50
microfiches, Belgian patents, per microfiche urgent despatch	100
aperture cards, Belgian patents, per patent	100
certified true copies, supplement per patent (Belgian or European)	100

24

Postage is included except for the *Recueil* and for air mail despatch.

Self service copying equipment is available in the reading room at a cost of 5 BF per page.

To reduce delays it is recommended that a deposit account be opened with OPRI (compte 000 – 2005878 – 15) mentioning *"ouverture d'un compte courant copies"*. A credit note is sent by post with the number of the deposit account to be quoted on all orders for copies.

V   Register of legal status

Bibliographic data and details of stages of progress are entered in the Registers, alphabetical and numerical, about one month after the relevant date. It is proposed to provide copies at 15 or 30 BF per page of extracts from the Registers.

OPRI can provide a written statement relating to the legal position of a specified Belgian patent or European patent designating Belgium for a fee of 500 BF. Requests for this service must be in writing and payment made by fiscal stamps or cheque in BF drawn on a Belgian bank.

VI Public services

Library
Open Monday – Friday 9.00 – 12.00 and 13.00 – 16.00

Patent documentation may be consulted in the reading room by any member of the public without charge.

Belgian, European, PCT and foreign patent documents are arranged in numerical order. There are two card files for Belgian patents, classified from 1950 (by IPC to sub-class level) and alphabetical from 1900 by name of applicant (individual persons and firms). The IPC classified lists are published in the relevant issue of the *Recueil*.

Abstracts of European and PCT patent documents are available on paper and on microfilm arranged by IPC (to the finest divisions).

It is proposed to provide on request classified lists of Belgian patents and applications from 1973 and European patents effective or having been effective in Belgium.

Patent specifications and official gazettes held in the library of the Belgian patent Office

AT	Austria	specns	1926 –
		gazette	1927 –
AU	Australia	specns	1904 –
		gazette	1905 –
BE	Belgium	specns	1830 –
		gazette	1854 –
CA	Canada	abridgments	1872 –
		gazette	1975 –
CH	Switzerland	specns	1888 –
DD	German Democratic Republic	specns	1974 –
		gazette	1984 –
DE	Germany, Federal Republic of	specns	1877 –
		gazette	1877 –
DK	Denmark	indexes	1927 –
EP	European Patent Office	specns	1978 –
		gazette	1978 –
ES	Spain	specns	1986 –
FI	Finland	gazette	1972 –
FR	France	specns	1791 –
		gazette	1884 –
GB	United Kingdom	specns	1617 –
		gazette	1854 –
IL	Israel	abridg. in English	1973 –
IT	Italy	specns	1926 –
		gazette	1919 –
JP	Japan	Pat. Abstr. of JP	1971 –
LU	Luxembourg	gazette	1910 – 72
NL	Netherlands	specns	1913 –
		gazette	1924 –
NO	Norway	specns	1974 –
		gazette	1933 – 73
NZ	New Zealand	gazette	1953 – 74; 1986 –
SE	Sweden	specns	1885 –
		gazette	1971 –
SU	Soviet Union	abridg. in English	1959; 1961 – 83
US	United States	specns	1861 –
		gazette	1872 –
WO	WIPO-PCT	specns	1978 –
		gazette	1978 –

Database

This database contains details of Belgian patent applications and granted patents and European patents effective or having been effective in Belgium. It is accessible in the library where the terminal is operated by OPRI personnel.

- Consultation of the register is free
- Printed copy of the entry in the Register is provided at the cost of 15 or 30 BF per page
- Provision proposed of specialized information, e.g. lists of one or two classifications; tariff to be decided.
- Online host Belindis (from 1987 at latest); tariff to be decided. Contact M Lauwerys, 30 rue J A De Mot, 1040 Brussels. Tel: 2 – 233 66 96 Telex: 23509 ENERGI B

Data accessible

- Applications 1973 – 83 (old law)
    Number, dates of application and grant, applicant, title, priorities, etc.
    Abstract (Belgian patents); incomplete before 1983.
    IPC, limited to subclass for Belgian patents.
- Applications from 1987
    All bibliographic data and stages of progress.

It is proposed to form at Brussels a team of engineers trained in the use of foreign databases specialized in patents. This facility is likely to be useful to small and medium sized enterprises, small research groups and independent inventors and technicians.

BUREAU GEVERS   **BE**
7 rue de Livourne
B – 1050 Bruxelles
Belgium
Tel: (02) 538 91 80
Telex: Brux. 26.407

Revue Gevers des Brevets   monthly   1956 –

This publication provides bibliographic details of Belgian patents in which the
entries are arranged by IPC and numerically within each class. There is also an
alphabetical name index of applicants and a bibliography of articles on industrial property. The preface to the journal is in French, Dutch, English and
German.

Public services

The *Bureau Gevers* offers an information and search service (Service de
Documentation technique) and has online access to the principal databases concerned with patent publications, e.g. WPI/WPIL, INPADOC, CAS, JAPIO,
INPI, CLAIMS, etc. There is also a database covering Belgian patents filed
from 1945 and patents of Zaire from 1950 – 72.

The services available include: –
– the provision of documentation related to a particular technical field
– the provision of lists of patents and patent applications in a particular
  technical field with respect to the principal industrial countries from 1974.
  Advice can be given on novelty, inventive activity and validity as well as pro-
  spects for exploitation.
– technical surveillance
– provision of copies of documents cited

PATENT OFFICE

BUNDESAMT FÜR GEISTIGES EIGENTUM (BAGE)
Einsteinstraße 2
CH – 3003 Bern
Tel. (031) 61 41 11
Telex. 912805 bage ch

I   General information

The Patent Office is a sub-division of the Federal Swiss Justice and Police Department and is responsible for patents, trade marks, honorary distinctions, indications of source or appellations of origin and markings, industrial designs and models and the law of copyright.

Staff (total):   180
Patent examiners: 55

II   Patent documents

Since 1954 patent applications in the fields of chronometry and textile fibres have been subjected to full substantive examination. These specifications are published (A3) for opposition after examination and again (B5) after grant. The specifications of other applications are published (A5) after grant.

The official languages are German, French and Italian though most are in German and a very few in Italian. Those published before grant are entitled *Auslegeschrift – Fascicule de la demande* and after grant *Patentschrift – Fascicule du brevet.*

III   Official gazette and other publications

Patentliste 1888 – 1961 twice a month
Lists of granted patents with bibliographic details.

SCHWEIZERISCHE EIDGENOSSENSCHAFT
BUNDESAMT FÜR GEISTIGES EIGENTUM

(19)

(11) **CH 654 969** A5

(51) Int. Cl.⁴: **H 05 K** 7/14

Erfindungspatent für die Schweiz und Liechtenstein
Schweizerisch-liechtensteinischer Patentschutzvertrag vom 22. Dezember 1978

(12) **PATENTSCHRIFT** A5

(21) Gesuchsnummer:	3764/81	(73) Inhaber: Rose-Elektrotechnik GmbH & Co. KG, Porta Westfalica (DE)
(22) Anmeldungsdatum:	09.06.1981	
(30) Priorität(en):	10.06.1980 DE 3021655	(72) Erfinder: Haseke, Horst, Porta Westfalica (DE)
(24) Patent erteilt:	14.03.1986	
(45) Patentschrift veröffentlicht:	14.03.1986	(74) Vertreter: A. Braun, Braun, Héritier, Eschmann AG, Patentanwälte, Basel

(54) Vorrichtung zum Befestigen von Leiterkarten oder Montagewänden

(57) Die Befestigung von Leiterkarten oder Montagewänden innerhalb eines Elektronik-Schaltgehäuses (1) erfolgt mittels Klemmvorrichtungen (4), durch die aus Längs- und Querstreben (2, 3) aufgebaute Montagerahmen im Schaltgehäuse (1) fixiert werden. Die Klemmvorrichtung (4) enthält eine Aufnahmenut (9), in welcher die Leiterkarten oder Montagewände durch Schraubmittel (10) an einer Halteschiene (6) festgehalten sind. Die Klemmvorrichtung (4) ist ihrerseits durch eine Formfeder (7) in einander gegenüberliegenden mit Nuten versehenen Seitenwänden des Gehäuses verankert. Die Formfeder (7) besitzt eine wellenförmige Gestalt und erfährt beim Niederdrücken der Federwölbung eine Verlängerung bezüglich der Länge der Halteschiene (6), wobei die Federenden in die Seitenwandnuten eingreifen.

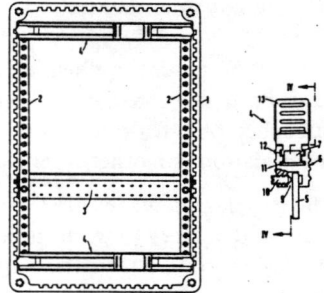

30

Schweiz. Patent-, Muster- und Markenblatt (PMMBl)   1962 –
In five parts.

Contents

Jahreskatalog der Patente, Muster und Modelle   1888/89 –
Annual name and class index of patents, designs and models.

IV   Sources of supply and prices (from 1 January 1986)

	SFr.
Specifications, *Patentschriften, Auslegeschriften*	9.00
By subscription for whole sections, classes or	
groups of IPC (add 0.30 for overseas)	8.10
Photocopies of foreign specifications	
per page (minimum 9.00)	0.50

Official gazette – *PMMBl*
By subscription

Section A (Parts I, II, III)		194.00
	overseas	(199.00)
Section B (Parts I, IV)		177.00
	overseas	(182.00)
Section C (Parts I – V)		341.00
	overseas	(353.00)
Section D (Part V)		41.00
	overseas	(46.00)
Part I as a supplement to subscriptions for Sections A, B, C or D		41.00
	overseas	(46.00)
Separate subscription for Part I		51.00
	overseas	(58.00)
Annual index to granted patents, designs and models		250.00
	overseas	(255.00)

Obtainable from: –

Bundesamt für geistiges Eigentum
Einsteinstraße 2
CH – 3003 Bern

V Register of legal status

This contains bibliographic data on patent specifications from the date of the first publication in 1888 with details of legal status. The entries are in numerical order and there is a fee for consultation.

VI Public services

Library

This is open to the public for the consultation of books and periodicals on industrial property matters as well as indexes.

Patent Office reading room (*Lesesaal*)

Open: Monday to Friday 8.00 – 11.45 and 13.30 – 17.00
This reading room containing patent literature is primarily for staff but the public may also make use of the material.

AP	ESARIPO	gazette	1984 –
AT	Austria	specns	1894 –
		gazette	Teil I 1955 –
			Teil II last 20 years
AU	Australia	specns	1956 –
		gazette	last two years
BE	Belgium	specns	1950 –
BG	Bulgaria	specns	1962 –
		gazette	last two years
BR	Brazil	gazette	last two years
CA	Canada	specns	1968 –
		gazette	last two years
CH	Switzerland	specns	1888 –
		gazette	1962 –
CN	China	specns	1985 –
		gazette	1985 –
CS	Czechoslovakia	specns	1919 –
		gazette	last two years
DD	German Democratic Republic	specns	1974 –
		gazette	last 20 years
DE	Germany, Federal Republic of	specns	1877 –
		gazette	1894 –
DK	Denmark	specns	1973 –
		gazette	last two years
EP	European Patent Office	specns	1978 –
		gazette	1978 –
ES	Spain	specns	1979 –
		gazette	last 20 years
FI	Finland	specns	1975 –
		gazette	last two years
Fr	France	specns	1920 –
		gazette	current year
GB	United Kingdom	specns	1945 –
		gazette	last two years
HU	Hungary	specns	1896 –
		gazette	last two years
IL	Israel	gazette	last two years
IT	Italy	specns	1928 –
		gazette	last 20 years
JP	Japan	specns	1971 –
		Pat.Abstr.of JP	1977 –
KR	Korea	abstracts	1979 –
LU	Luxembourg	gazette	last two years
MX	Mexico	gazette	last two years
NL	Netherlands	specns	1913 –
		gazette	last 20 years
NO	Norway	specns	1974 –
		gazette	last two years
NZ	New Zealand	specns	1981 –

OA	OAPI	specns	1966 –
		gazette	1966 –
PL	Poland	specns	1924 –
		gazette	last two years
PT	Portugal	gazette	last two years
RO	Romania	specns	1957 –
SE	Sweden	specns	1972 –
		gazette	last 20 years
SU	Soviet Union	specns	1954 –
		gazette	last one year
US	United States	specns	1920 –
		gazette	1945 –
WO	WIPO-PCT	specns	1978 –
		gazette	1978 –
YU	Yugoslavia	specns	1961 –
		gazette	last two years

Zentrale Patentschriftensammlung (ZPS)

This is a collection of specifications arranged by IPC which is located in the Patent Office reading room and freely accessible to the public. It contains: –

AT Austria 1975 – ; CA Canada 1983 – ; CH Switzerland 1969 – ; DE Germany, Fed. Rep. 1975 – ; EP European Patent Office 1978 – ; FR France (*Abrégés*) 1975 – ; GB United Kingdom 1975 – ; NL Netherlands 1975 – ; US United States 1975 – ; WO PCT 1978 – .

Special services

Since 1984 the Patent Office has been able to offer a state of the art service on patents known as TIPAT, Technical Information on Patents. This enables a potential patentee to obtain quickly information on the subject field of his invention and on known solutions to a particular technical problem.

Up to date search equipment is available for this work with access to the principal commercial data banks as well as those of the European Patent Office and INPADOC.

Requests should be in writing indicating precisely the required scope of the search. Fees payable are based on the time taken by the researcher and the cost of data bank consultations.

VII Provincial libraries

Swiss patent documents may be consulted by the public free of charge at the following addresses; European and PCT also at those libraries indicated*. (Cl in classified order; Nr in numerical order)

* BASEL(Cl)

Gewerbemuseum
Öffentliche Fachbibliothek
Spalenvorstadt 2

Open Tu 10 – 12, 14 – 18; Wed 10 – 12, 14 – 21; Th 10 – 12, 14 – 18; Fr and Sat 10 – 12, 14 – 17.

LA CHAUX-DE-FONDS (Nr)

Technicum neuchatelois
rue du Progrès 40

Open Mo – Fr 9 – 12; 14 – 17.

CHUR (Nr)

Gewerbebibliothek
Scalettastrasse 33

Open Tu and Th 19.30 – 21.30.

*GENEVA (Cl)

Collection des brevets suisses
avenue du Cardinal-Mermillod 5
Carouge

Open Mo – Fr 9 – 12, 14 – 17.

LUGANO (Nr)

Archivo comunale
Citta di Lugano
Corso Elvezia 36

Open Mo – Fr 9 – 12, 14 – 18.

*WIL SG (Cl)

Patentbibliothek
Werkstrasse 1

Open Tu 14 – 17; Th 9 – 11.30; 14 – 17.

*ZÜRICH (Cl)

Patentschriftensammlung
Vorderberg 11

Open Tu, Wed and Th from 14.00.

PATENT DOCUMENTATION GROUP   CH

Secretary   Dr P Ochsenbein
            c/o CIBA-GEIGY AG
            R – 1046.4.02
            PO Box 2543
            CH – 4002
            Basel
            Switzerland
Tel: 061 26 50 08

A working pool of thirteen chemical and oil companies was formed in 1957 for the purpose of abstracting patent documents and searching for equivalent patents. When the publications of Derwent began to meet the requirements of industrial documentalists the PDG was restructured in 1969 in order to improve both its organization and representation.

Membership

In 1984 the Statutes of the PDG were changed in order to allow non-chemical companies to become members. The current membership is 23 companies in five countries, viz. Belgium (1), France (3), Germany (9), Netherlands (7), Switzerland (3). Membership fee is 3400 Swiss Francs.

Objectives

The statutory purpose of the PDG is "to provide for cooperation in information and documentation from and for patents" and "to exchange knowledge and experience" in various areas.

The objectives within this framework are to discuss problems, reveal common interests, prepare proposals and to voice them to the public and in particular to authorities and other organizations active in the field of intellectual property, e.g. to: –

- national and international associations and institutions representing users of patents and patent information
- commercial and governmental services assembling and disseminating patents and patent information
- national and supranational patent offices as producers of patents and patent information
- WIPO as an intergovernmental organization promoting patent protection and disseminating patent information

Activities

There are several Working Groups in which specialists submit observations and data concerning new or changed patent information and documentation media and systems. Where considered necessary proposals are prepared to overcome errors and shortcomings as well as advice regarding trends and developments. The main areas of this activity are: –

– hardware and software characteristics of electronic databases, search profiles, search tools of all kinds, networks
– physical characteristics, layout, presentation, data elements, coding, etc., of and in printed patent specifications, official gazettes and indexes and in microfilmed or computerized files thereof, produced by official authorities or commercial services
– technical and organizational developments in the application of library automation
– trends and user requirements in connection with sophisticated systems of computerizing chemical formulae

PATENT OFFICE

DEUTSCHES PATENTAMT
Zweibrückenstraße 12
D – 8000 Munich 2
Tel: (089) 21 95 1
Telex. (05) 23534 BPBM D
Telecopier 21 95 22 21

DEUTSCHES PATENTAMT
– Dienststelle Berlin –
Gitschiner Straße 97 – 103
D – *1000 Berlin 61*
Tel. (030) 2 59 40
Telex. (018) 3604 D

I   General information

The German Patent Office is an administrative body under the *Bundesminister der Justiz* (Federal Minister of Justice). It is an examining patent office and responsible for the granting of patents and the registration of utility models (*Gebrauchsmuster*), trade marks and service marks in the Federal Republic of Germany. This work is done in the Munich office. The branch in Berlin is responsible for the sale of documents and for the administration of German trade marks registered up to the end of the war. In Berlin also patent applications, oppositions, etc may be filed and fees paid. Both Munich and Berlin can provide information and have libraries containing material in technical and industrial property fields as well as classified collections of patent documents.

Staff (total) Munich and Berlin: 2250
Patent examiners (Munich):   620
Library staff (Munich and Berlin):   100

⑲ BUNDESREPUBLIK   ⑫ **Patentschrift**

    DEUTSCHLAND   ⑪ **DE   32 46 683   C 1**

⑤ Int. Cl. ³:

C 21 D 9/08

C 21 D 1/63

DE 32 46 683 C 1

**DEUTSCHES
PATENTAMT**

㉑ Aktenzeichen: P 32 46 683.8-24
㉒ Anmeldetag: 14. 12. 82
㊸ Offenlegungstag: —
㊺ Veröffentlichungstag
    der Patenterteilung: 15. 3. 84

Innerhalb von 3 Monaten nach Veröffentlichung der Erteilung kann Einspruch erhoben werden

㉓ Patentinhaber:

Mannesmann AG, 4000 Düsseldorf, DE

㉒ Erfinder:

Völlmecke, Hermann G., Dipl.-Ing., 4330 Mülheim,
DE; Hillemanns, Herbert, Dipl.-Ing., 4030 Ratingen,
DE; Ribken, Hans, Ing.(grad.), 4006 Erkrath, DE;
Carneim, Wilfried, Dipl.-Ing., 4018 Langenfeld, DE;
Homberg, Gerd, Dr.-Ing., 4021 Metzkausen, DE

㊽ Im Prüfungsverfahren entgegengehaltene
Druckschriften nach § 44 PatG:

DE-PS   29 35 242

DE 32 46 683 C 1

㊼ Druckentlastungseinrichtung für eine Rohr-Ölvergütungsanlage

Die Erfindung betrifft eine Druckentlastungseinrichtung für
eine Rohr-Ölvergütungsanlage, bestehend aus einer mit Öl
gefüllten Wanne, sowie Zufuhr- und Austragseinrichtungen für
die Rohre sowie einer im unteren Teil der Wanne angeordne-
ten Haltevorrichtung für die Rohre mit einer zugeordneten
Düse zum Beaufschlagen des Rohrinneren mit Öl, bei der
eine die Öffnung der Wanne weitgehend überdeckende in das
Ölbad eintauchende Abdeckhaube vorgesehen ist, deren der
Zufuhreinrichtung naheliegende Wand kürzer ist als die der
Austragseinrichtung naheliegende Wand. Um das Auftreten
eines starken Überdrucks unter der Abdeckhaube infolge der
Bildung von Öldämpfen und Verbrennungsgasen, die durch
das Eintauchen der glühenden Rohre in das Ölbad entstehen,
wirkungsvoll zu verhindern, wird erfindungsgemäß vorge-
schlagen, daß sich auf der Abdeckhaube (4) ein blasebalg-
artiger Körper (1) befindet, der mit mindestens einer An-
schlußleitung (2) ausgestattet ist, deren Öffnung (3) in dem
Sammelraum endet.   (32 46 683)

Fig. 1

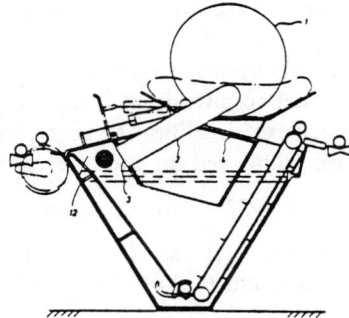

BUNDESDRUCKEREI   01. 84   408 111/358   80

39

II Patent documents

In accordance with the current patent law (of 1. 1. 81) the following documents are, or have been, published: −

Offenlegungsschrift (OS) (A)
The unexamined application laid open to public inspection and published 18 months after the application or priority date, (on yellow paper, aperture cards with yellow edge).

Issued since Oct. 1968 (No. 1 400 001)

Auslegeschrift (AS) (B)
Specification of examined and accepted application published to allow opposition by third parties, (on green paper, aperture cards with green edge). Following the latest patent law amendment these are no longer published.

1955 − 56   Published as the so called *"weiße Auslegeschriften"*.
1957 − 81   With a few continuing up to 1985.

Patentschrift (PS) (C)
Specification of the granted patent, (on white paper, aperture cards without coloured edge).
1877 − 1945   1 − 768 160   Printed patent specifications were issued. Filing documents, published applications and patent specifications from the last years of the war, about 190,000 applications without continuous numbering, are on microfilm in the FIAT-Berichte (Field Intelligence Agency Technical Reports). The films are stored in Munich; there is a classified collection of abstracts as a searching aid.

1949 −   800 001 −   Printed patent specifications are issued.

Gebrauchsmuster (Gbm) (U)
Printed specifications were not published before 1968 but are available in the Patent Office libraries as *Gebrauchsmusterunterlagen*. Since 1968 they have been published on aperture cards.

Before 1933   No copies exist.

1933 − 45   (1 286 500 − 1 534 499)   The documents are available on film in Munich and Berlin. About 5000 numbers between 1943 and 1945 are missing.

1949 − 68   (1 600 001 − 1 994 700)   Complete collections available in Munich and Berlin.

1968 −   (1 994 − 1 999 000 and 66 00 001 −   ) Complete collections in the form of aperture cards are available in Munich and Berlin.

Translations of PCT applications

If published PCT applications designating the Federal Republic of Germany are not in the German language the German Patent Office publishes translations (on pink paper, aperture cards with pink edge). Filing and publication numbers in the 90 000 range have been reserved.

Translations of EP patent claims

At the request of the applicant EP applications designating the Federal Republic of Germany the German Patent Office publishes translations of the claims (on light blue paper).

III Official gazette and other publications

Patentblatt 1877 – weekly

Inhaltsverzeichnis	Table of Contents
Teil 1: Offengelegte Patent-anmeldungen	Part 1: Patent applications published prior to examination
a) Offenlegungen	a) Publications prior to examination
b) Eingang von Anträgen auf Neuheitsrecherche	b) Requests for novelty search received
c) Mitteilung von Recherchen-ergebnissen	c) Notification of search results
d) Eingang von Prüfungsanträgen	d) Requests for examination received
e) Unwirksamkeit von Prüfungs-anträgen	e) Ineffective requests for examination
f) Lizenzbereitschaft vor der Bekanntmachung	f) Declarations of willingness to grant grant licenses, made prior to publication
g) Änderungen	g) Changes
h) Zurücknahmen, Zurückweisungen und sonstige Erledigungen	h) Applications withdrawn, rejected or otherwise disposed of
i) Verschiedenes	i) Miscellaneous
Teil 2: Bekanntgemachte Patentan-meldungen	Part 2: Patent applications published after examination
a) Bekanntmachungen	a) Publications after examination
b) bis e) frei	b-e) Vacant
f) Lizenzbereitschaft	f) Declarations of willingness to licenses
g) Änderungen	g) Changes
h) Zurücknahmen, Zurückweisungen und sonstige Erledigungen	h) Applications withdrawn, rejected or otherwise disposed of
i) Verschiedenes	i) Miscellaneous

Teil 3: Erteilte Patente	Part 3: Patents granted

Teil 3: Erteilte Patente	Part 3: Patents granted
a) Erteilungen	a) Grants
b) bis e) frei	b-e) Vacant
f) Lizenzbereitschaft	f) Declarations of willingness to grant licenses
g) Änderungen	g) Changes
h) Löschungen	h) Cancellations
i) Verschiedenes	i) Miscellaneous
Teil 4: Gebrauchsmuster	Part 4: Utility models
a) Eintragungen	a) Registrations
b) Verlängerung der Schutzdauer	b) Extensions of the period of protection
c) bis f) frei	c-f) Vacant
g) Änderungen	g) Changes
h) Löschungen	h) Cancellations
i) Verschiedenes	i) Miscellancous
Teil 5: Europäische Anmeldungen und Patente mit Benennung der Bundesrepublik Deutschland	Part 5: European applications and patents designating Germany
a) Veröffentlichte europäische Anmeldungen	a) Published European applications
b) Veröffentlichte Übersetzungen der Patentansprüche europäischer Anmeldungen	b) Published translations of claims of European applications
c) Erteilte europäische Patente	c) Granted European patents
d) bis g) frei	d-g) Vacant
h) Erledigungen ohne Patenterteilung	h) Cancelled without grant
i) Verschiedenes	i) Miscellaneous
Teil 6: Internationale Anmeldungen (PCT)	Part 6: International applications (PCT)
a) Internationale Veröffentlichungen in deutscher Sprache	a) Published applications in German
b) Internationale Veröffentlichungen in deutscher Übersetzung	b) Published applications translated into German
c) bis g) frei	c-g) Vacant
h) Zurücknahmen und sonstige Erledigungen	h) Withdrawn and cancelled
i) Verschiedenes	i) Miscellaneous

Each issue has a name index covering all six sections and a serial number to IPC concordance for each section.

Vierteljährliches Namensverzeichnis zum Patentblatt 1895 – Quarterly

Alphabetical list of applicants (not inventors) with references to the relevant entries in the *Patentblatt*.

Jahresverzeichnis der Auslegeschriften und erteilten Patente 1957 – Annual

Annual index to *Auslegeschriften* and granted patents.

Inhaltsverzeichnis	Table of Contents
1. Nummernübersicht der Patente	1. Survey in numerical order of patents granted which have been published in the Patentblatt during the year, indicating the classification.
2. Gruppenverzeichnis der DAS und Patente	2. Classified list of "Auslegeschriften" issued during the year and patents granted which have been published in the Patentblatt during that year.
3. Noch nicht gelöschte Patente	3. List of patents in force.
4. Verzeichnis der im Jahre 19 . . für nichtig und teilweise für nichtig erklärten Patente und der beschränkten Patente	4. Index of patents declared invalid or partially invalid and of patents limited.

IV Sources of supply and prices

– All German documents and aperture cards as well as photocopies of Austrian and Swiss patent documents and of French documents from No 2 055 001 from:

DEUTSCHES PATENTAMT, DIENSTSTELLE BERLIN
– Schriftenvertrieb –
Gitschiner Straße 97 – 103
D-1000 Berlin 61

and some commercial publishers.

– Photocopies of other foreign patent documents from:

DEUTSCHES PATENTAMT MÜNCHEN
– Lichtbildstelle –
Zweibrückenstraße 12
D-8000 München 2

– Journals from:

CARL HEYMANN VERLAG KG
Gereonstraße 18 – 32
D-5000 Köln 1

– Aperture cards of *OS, AS* and *PS* from:

DEUTSCHER PATENT-DIENST GMBH
Döllingerstraße 17
D-8000 München 19

and

LABOVO
Patentwirtschaftsdienst
Würmseestraße 59
D-8000 München 71

	Prices from 1. Jan. 1985
	DM
German patent documents (*OS, AS, PS, Gbm*)	
per document	5.50
by subscription to entire IPC-sections, classes, groups or sub-groups per document	5.20
Aperture cards (*OS, AS, PS, Gbm*)	
per document	3.10
by subscription to entire IPC-sections, classes, groups or sub-groups per document	2.80
Photocopies of foreign documents, per page*	1.00
Patentblatt, quarterly subscription rate	170.00
Vierteljährliches Namensverzeichnis, yearly subscription rate	140.00
Jahresverzeichnis der Auslegeschriften und erteilten Patente	174.00

Documents are despatched within a few days when ordered by post and usually the same day when ordered by telephone or telex. There may be an additional delay of two days if printed documents are no longer available and copies are supplied from aperture cards. Documents ordered must be paid for on delivery or through an account at the *Schriftenvertrieb* in Berlin and/or the *Lichtbildstelle* in Munich.

For further details see:

Merkblatt über den Bezug von Druckschriften, Filmlochkarten und anderen Unterlagen des Deutschen Patentamts sowie Einsichtnahme in diese Unterlagen (Ausgabe Januar 1985)

---

*) Several commercial firms operating within the Patent Office building in Munich can supply photocopies at DM 0.60 per page. See also self-service on para. VI

V   Register of legal status

Bibliographic data and stages of progress of all German patent documents, OS, AS and PS from 1877 are stored on the *Patentrolle.* This register is arranged numerically and there is free public access. Telephone enquiries are accepted. The same information for *Gebrauchsmuster* is stored on the *Gebrauchsmusterrolle,* also arranged numerically and similarly accessible.

Telephone enquiries

Munich	(0 89) 21 95-22 91, 22 92 and 22 93 for patents
	(0 89) 21 95-24 45 for *Gebrauchsmuster*
Berlin	(0 30) 25 94-691, 692

Since Autumn 1984 the German Patent Office has made access to the computerized register possible via external terminals and telex. Further information on this may be obtained from the Patent Office.

VI   Public services

The public libraries of the German Patent offices are open:
Monday – Thursday 7.30 – 16.00 and Friday 7.30 – 14.45

Access to the library in both Munich and Berlin is through the public search room.

Holdings include classified lists to IPC, DPK (the German system of classification) and the US system as well as search file lists to DPK/IPC for United States, Great Britain and France from 1954.

From 1950 to 1975 there is a name index and a classified index for *Auslegechriften* and *Gebrauchsmuster*. This is continued on INPADOC.

Abstracts of *Gebrauchsmuster* are arranged in a card index from 1964, to 1973 by DPK and from 1974 by IPC.

There are two self service photocopy machines in the library at Munich (DM 0.30 per page).

There are the following classified collections at Munich:

Austria	1974 –	United Kingdom	1954 –
France	1954 –	USA	1954 –
Germany (Dem. Rep.)	1977 –	EPO	1978 –
Germany (Fed. Rep.)	1877 –	PCT	1978 –

In Berlin there are: France 1973 – , German Dem. Rep. 1977 – , Germany, Fed. Rep. 1877 – , EPO and PCT.

Patent specifications and official gazettes held in the library of the German Patent Office in Munich

AP	ARIPO	gazette	1984 –
AR	Argentina	gazette	1971 –
AT	Austria	specns	1899 – 1943; 1949
		gazette	1899 – 1941; 1949 –
AU	Australia	specns	1939 –
		gazette	1931 –
BE	Belgium	specns	1950 –
		gazette	1928 –
BG	Bulgaria	specns	1962 –
		gazette	1962 –
BR	Brazil	specns	1971 –
		gazette	1972 –
CA	Canada	specns	1962 –
		gazette	1928 –
CH	Switzerland	specns	1888 –
		gazette	1889 –
CN	China	specns	1985 –
		gazette	1985 –
CO	Colombia	gazette	1958 –
CS	Czechoslovakia	specns	1938 –
		gazette	1945 –
DD	German Democratic Republic	specns	1951 –
		gazette	1952 –
DE	Germany, Federal Republic of	specns	1877 –
		gazette	1877 –
		Gbm	1949 –
DK	Denmark	specns	1920 –
		gazette	1920 –
EG	Egypt	specns	1955 – (incomplete)
		gazette	1951 – 77
EP	European Patent Office	specns	1978 –
		gazette	1978 –
ES	Spain	abstracts	1984 –
		gazette	1953 –
FI	Finland	specns	1944 –
		gazette	1945 –
FR	France	specns	1791 –
		gazette	1884 –
GB	United Kingdom	specns	1617 –
		gazette	1919 –
GR	Greece	gazette	1966 –
HU	Hungary	specns	1947 –
		gazette	1896 – 1940; 1945 –

IE	Ireland	gazette	1960 –
IL	Israel	gazette	1962 –
IN	India	specns	1945 –
		gazette	1947 –
IT	Italy	specns	1935 –
		gazette	1902 – 36; 1943 –
JP	Japan	specns	1950 –
		gazette	1950 –
		Pat. Abstr. of JP	1977 –
LU	Luxembourg	gazette	1945 –
NL	Netherlands	specns	1913 –
		gazette	1912 –
NO	Norway	specns	1945 –
		gazette	1946 –
OA	OAPI	specns	1980 –
		gazette	1966 – 70; 1977 –
PH	Philippines	gazette	1962 –
PL	Poland	specns	1945 –
		gazette	1959 –
PT	Portugal	gazette	1937 – 41; 1950 –
RO	Romania	specns	1957 –
		gazette	1961 –
SE	Sweden	specns	1885 –
		gazette	1905 – 19; 1946 –
SU	Soviet Union	specns	1924 –
		gazette	1958 –
TR	Turkey	gazette	1946 –
US	United States of America	specns	1867 –
		gazette	1890 –
WO	WIPO-PCT	specns	1978 –
		gazette	1978 –
YU	Yugoslavia	specns	1961 –
		gazette	1951 –

A comprehensive collection of books and periodicals on scientific, technical and legal subjects is available in the public search room.

Patent specifications and official gazettes held in the library of the German Patent Office in Berlin

AT	Austria	specns	1899 –
		gazette	1936 –
AU	Australia	specns	1939 –
		gazette	1931 –
BE	Belgium	specns	1978 –
		gazette	1932 –

BG	Bulgaria	specns	1955 –
BR	Brazil	specns	1971 –
		gazette	1971 –
CA	Canada	specns	1970 –
		gazette	1939; 1949 –
CH	Switzerland	specns	1888 –
		gazette	1962 –
CO	Colombia	gazette	1950 –
CS	Czechoslovakia	specns	1927 –
		gazette	1968 –
CU	Cuba	gazette	1967 –
DD	German Democratic Republic	specns	1976 –
		gazette	1960 –
DE	Germany, Federal Republic of	specns	1877 –
		gazette	1877 –
		Gbm	1933 –
DK	Denmark	specns	1895 – 1945; 1971 –
		gazette	1937 –
FI	Finland	specns	1944 –
FR	France	specns	1791 –
		gazette	1941 –
GB	United Kingdom	specns	1631 – 1915; 1972 –
		gazette	1915 – 19; 1943 –
		abstracts	1855 –
HU	Hungary	gazette	1900 – 44; 1962 –
IT	Italy	gazette	1913 – 19; 1924 – 39; 1949 – 54.
JP	Japan	specns	1952 –
		Pat. Abstr. of JP	1977 –
LU	Luxembourg	gazette	1961 –
NL	Netherlands	specns	1941 –
		gazette	1920 –
NO	Norway	specns	1946 –
		gazette	1911 – 19; 1931 – 39; 1944 –
PL	Poland	specns	1974 –
		gazette	1972 –
SE	Sweden	specns	1951 –
		gazette	1945 –
SU	Soviet Union	specns	1960 –
		gazette	1959 –
US	United States of America	specns	1956 –
		gazette	1907 –
YU	Yugoslavia	specns	1922 –

A comprehensive collection of books and periodicals on scientific, technical and legal subjects is available in the public search room.

The following INPADOC services are available in the reading rooms:

Numerical Data Base (NDB)
Patent Classification Service (PCS)
Patent Inventor Service (PIS)
Patent Applicant Service (PAS)
INPADOC Patent Gazette (IPG)
Concordances

Online searches of the INPADOC Patent Family Service may be made in the library with or without information on legal status, at a cost of DM 230 or DM 150 per family searched.

The computerized patent and *Gebrauchsmuster* indexes can be used free of charge in the reading rooms. Access from outside is also possible at a cost of DM 1 for each number searched.

Books from 1975 are also classified to IPC and may be searched at the computer terminals in the library.

The database RALF (*Rechtstands-Auskunft und Lizenzforderungsdienst* – legal status and licence promotion service) provides up to date information on "licence of right" declarations. This service is provided by the Berlin branch and online searches may be made in the public search rooms in Munich and Berlin. The possibility of external online searching is being considered.

Prices: Profile search, title page – DM 1 for each document identified and title page despatched; Profile search, single retrospective – DM 0.50 for each identified document; Profile search, legal status – DM 0.20 for each file number with brief information; Single search – DM 1 for each section of the IPC searched and DM 0.50 for each identified document.

As required by para. 29, section 3 of the Patents Act, the German Patent Office, for a fee of DM 850, will carry out research on technical problems, even if not associated with patent protection. The research report lists relevant documents but without an assessment of their importance. Photocopies are provided.

There is also a monitoring service for the identification of publications of special interest. This service covers German patent specifications, *Gebrauchsmuster* and European and PCT specifications in which Germany has been designated.

Patent databases

PATDPA

The German patent database PATDPA contains the bibliographic data and abstracts of the *Offenlegungsschriften, Auslegeschriften* and *Patentschriften* from 1981 and is in the process of being backdated to 1968. The database, developed by the *Deutsches Patentamt,* the *Fachinformationszentrum Karlsruhe,* the *Satzrechenzentrum Berlin* and the *Gesellschaft für Information und Dokumentation,* will be extended into a full text database with reproduction of drawings. Access to PATDPA is through STN International.

Vendor: Fachinformationszentrum Energie, Physik, Mathematik GmbH
Postfach 2465, D-7500 Karlsruhe.
Tel.: 0 72 47/82-4600
Telex: 17724710+
Cost: DM 240 per hour of use, Online display DM 0.60

PATOS

The patent database PATOS contains the bibliographic data and the most significant claim of all *Offenlegungsschriften* since 1968 and also of those *Auslegeschriften* and *Patentschriften* for which there were no published *Offenlegungsschriften*. The text is derived from the abstracts published by WILA-Verlags and so far about one million documents are covered.

Vendor: Bertelsmann Informations Service GmbH
Neumarkter Straße 18, D-8000 Munich 80
Tel: 089-43189-0
Telex: 5 23 259 VBMUE

Cost per hour DM 320. Online display DM 2.50

VII Provincial libraries

Patentauslegestellen (PAS)

German patent documents are available to the public at the *Deutsches Patentamt* in Munich and Berlin and at twelve *Auslegestellen* in the following cities: –

Aachen	Darmstadt	Hamburg	Nürnberg
Bielefeld	Dortmund	Hannover	Ratingen
Bremen	Düsseldorf	Kaiserslautern	Stuttgart

The PAS are sponsored by various local authorities or professional organizations and hold complete collections with the exception of Ratingen which holds those classes relevant to the local industry.

The collections are arranged by DPK (German patent Classification) to the end of 1974 and thereafter by IPC. In addition to the holdings listed there are usually name and classified indexes and trade mark journals.

Most PAS have access to patent and general literature databanks and some impose an entrance fee. There is generally the facility of free consultation with a patent agent on certain days.

Abbreviations

OS	*Offenlegungsschriften*
AS	*Auslegeschriften*
PS	*Patentschriften*
AltPS	*Patentschriften* granted by the *Reichspatentamt*
DE-EP	Claims of EP patent applications designating DE in German translation
DE-WO	PCT specifications designating DE in German translation
Gbm	*Gebrauchsmuster*
Mitt	*Mitteilungen der Deutschen Patentanwälte*
BlPMZ	*Blatt für Patent-, Muster- und Zeichenwesen*
PatBl	*Patentblatt*
GRUR	*Gewerblicher Rechtsschutz und Urheberrecht*
GRURInt	*GRUR Internationaler Teil*

AACHEN

Bibliothek der Technischen Hochschule Aachen
Templergraben 61
D-5100 Aachen
Reading room – Jägerstraße 17 – 19
Tel. (0241) 80 44 80

Open: Mo – Fr 9.00 – 13.00
Fee: Daily DM 5.00, Annual DM 100.00

Patent documents

OS, AS, PS, AltPS, DE-EP, DE-WO: all classes in printed copy
EP: all classes on aperture card

Gbm: 1964 –
US *Official Gazette* (abstracts): 1881 –
GB Abstracts: 1979 –   Abridgments: 1929 – 82

Literature

*PatBl, BlPMZ, European Patent Bulletin, Official Journal of the EPO, Mitt,
GRUR, GRURInt*
Standards and Guidelines, DIN, VDI

Free consultation with patent agents for new inventors every second Wednesday
in the month 14.15 – 17.00

Online searches of literature and patent databanks available in the main library.

BIELEFELD

Stadtbibliothek Bielefeld
Herforder Straße, 4-6/Wilhelmstraße 3
D-4800 Bielefeld

Tel. (0521) 51 68 52

Open: Tu – Fr 9.00 – 16.00
        Sa   9.00 – 13.00

Patent documents
OS: selected classes; complete from 1984
AS: all classes from 1975
PS: all classes but since 1957 only PS differing from corresponding AS
Gbm: same selected classes as OS; complete from 1984

Literature
PatBl, BlPMZ, European Patent Bulletin, Official Journal of the EPO.

Standards and Guidelines, DIN, VDE

Free consultation with patent agents on the first Wednesday in the month,
16.00 – 18.00

BREMEN

Hochschule Bremen
Langemarckstraße 116
2800 Bremen 1

Tel. (0421) 59 05 225

Open: Mo – Fr 9.00 – 13.00

Patent documents
OS, AS, PS, AltPS: all classes, *Gbm,* DE-EP, DE-WO

Literature
GRUR, GRURInt

INPADOC: NDB

Free consultation with patent agents the first Thursday in the month,,
  15.30 – 17.00 at the Handelskammer, Hinter dem Schütting, 2800 Bremen 1.
Tel. (0421) 36 37 237

DARMSTADT

Hessische Landes- und Hochschulbibliothek
Schloß
D-6100 Darmstadt
Tel. (06151) 12 54 27

Open: Mo – Fr 8.00 – 16.00
  Every first and third Saturday in the month 8.00 – 12.00

Patent documents
OS: all classes, from 1971 on aperture card
AS: all classes in printed copy
PS: all classes in printed copy but only those PS differing from the cor-
     responding AS
AltPS: all classes in printed copy
Gbm: all classes, from 1972 on aperture card
AT 378 801 – ; DD abstracts 1978 – ; EP 1978 – , FR abstracts 1957 – , GB Der-
went abstracts 1967 – ; JP Derwent abstracts (Chem) 1967 – ; SU Derwent
abstracts 1962 – ; US abstracts 1950 – ; WO abstracts 1979 – .

Literature
PatBl, BlPMZ, European Patent Bulletin, Official Journal of the EPO.
GRUR, GRURInt, Chemical Abstracts

Standards and Guidelines: DIN, VDI, ASTM

INPADOC: NDB and PIS 1968 – 82

Online access to databanks FIZ Technik, INKA (for CAS)

Free consultation with patent agents on the first Tuesday in the month.

DORTMUND

Universitätsbibliothek
Vogelpothsweg 76
D-4600 Dortmund 50 (Barop)

Tel. (0231) 7 55 40 14
Telecopier.: (0231) 75 15 32

Open: Mo – Fr 9.00 – 16.00

Fee: Daily DM 5.00, Annual DM 100

Patent documents
OS, AS, PS, AltPS, Gbm, EP, DE-EP, DE-WO, WO
US Official Gazette 1872 – , *PCT Gazette*

Standards and Guidelines: DIN, VDI, VDE, ASTM

INPADOC: NDB, PAS, PIS, PCS, PFS (online)

Online access to registers at German and European patent offices and to patent and literature databanks

Free consultation with patent agents the first and third Wednesdays in the month, 14.00 – 16.00

DÜSSELDORF

VDI Dienstleistungen GmbH
Graf-Recke-Straße 84
4000 Düsseldorf 1

Tel. (0211) 6 21 42 00
Telex: 8 586 525
Telecopier: (0211) 6 21 45 75

Open: Mo, We, Th, Fr 9.00 – 16.00 Tu 9.00 – 12.00

Patent documents
OS, AS, PS, AltPS, Gbm, DE-EP, DE-WO, EP
US Official Gazette 1924 – ; SU Patent Bulletin

INPADOC: NDB

Online access to patent and literature databases

Free consultation with patent agents the first, second and third Wednesday in the month, 14.00 – 16.30.

HAMBURG

Handelskammer
IPC-Innovations- und Patent-Centrum
Börse
D-2000 Hamburg 11
Tel. (040) 36 13 83 76

Open: Mo – Fr 9.00 – 14.00

Patent documents
OS, AS, PS, AltPS, Gbm

Literature
PatBl, BlPMZ, Mitt

Online access to WPI, INPADOC, JAPIO, CLAIMS, PATDPA, PATOS, etc.

Free consultation with patent agents the first and third Thursday in the month, 14.00 – 15.00.

HANNOVER

Universitätsbibliothek der Technischen Universität
Hannover und Technische Informationsbibliothek
Lesesaal PIN (Patente, Informationen, Normen)
Welfengarten 1B
D-3000 Hannover 1
Tel. (0511) 762 3415
Telex. 922168 (tibhn d)
Telecopier. (0511) 715936

Open: Mo, We, Fr 9.00 – 16.30; Tu, Th 9.00 – 19.00; (univ. vac. 9.00 – 18.00)
Sa 9.00 – 12.00

No entry fees

Patent documents
OS (abstracts), *AS, PS, AltPS, Gbm* (abstracts)
Official gazettes: AT, CH, EP, GB, US
Abstracts: DD, EP, GB, SU, US

INPADOC: PIS, PAS

Online access to patent and literature databanks.

Free consultation with a patent agent at IHK Hannover-Hildesheim, Schiffgraben 49, 3000 Hannover on the first and third Wednesday in the month from 14.00 – 16.00. Appointments on (0511) 3107 275.

Inventor assistance through the Erfinderzentrum Norddeutschland, Friesenstraße 14, 3000 Hannover 1, Tel: (0511) 316058/316059

KAISERSLAUTERN

Universitätsbibliothek Patentschriftenauslegestelle
Erwin-Schrodinger-Straße
D-6750 Kaiserslautern

Tel. (0631) 205 2172

Open: Mo – Fr 8.00 – 16.30

No entry fee

Patent documents
OS, AS, PS, AltPS: 1880 – (from 1970 on aperture cards)
Gbm and EP on aperture cards from 1984.

Literature
PatBl, BlPMZ, European Patent Bulletin, Official Journal of the EPO
Literature on industrial law as well as books and periodicals on science and technology available in the nearby university library.

Standards DIN and VDI, *Chemical Abstracts.*

Access to numerous online databanks.

Free consultation with a patent agent on the first Wednesday in the month.

NÜRNBERG

Landesgewerbeanstalt Bayern
Abt. Information + Dokumentation
Marientorgraben 8
Gewerbemuseumsplatz 2
D-8500 Nürnberg 1

Tel.: (0911) 2 01 75 16
Telex: (06) 222 29
Telecopier: (0911) 20 17 504

Open: Mo – Th 9.00 – 16.00, Fr 9.00 – 15.00

Fees: Daily 10.00 DM, weekly 40.00 DM, monthly 150.00 DM, annual 1200.00 DM
No charge for unemployed persons, pensioners and students.

Patent documents
OS: Sept 1968 –
AS: Sept 1954 –
PS: 1877 –
Gbm: 1964 – 71 (abstracts); 1972 – (full specification)
    AT 273,900 – ; CH 1 – ; DD 1970 – (abstracts); FR 1959 – (abstracts)
    GB abstracts 1 – 600,000; 900,000 – ; US 1872 – (abstracts),
    3,226,729 – (on film), WO (abstracts)

Official gazettes: AT, CH, DD, DE, EP

INPADOC: NDB, PIS

Extensive stock of periodicals, books, standards and other non-patent literature.

Online access to Derwent, INPADOC, INPI, CLAIMS, PATOS, DPA 1, etc.

Free consultation with a patent agent on the first Wednesday in the month.

Special services

In addition the following search facilities are provided for medium and small industrial firms which may not have a patent department; fees according to LGA rates.
a) Searches covering several countries with respect to – state of the art and novelty, opposition procedures, equivalent patents, names.
b) Monitoring state of progress of individual applications.
c) Legal status of patent and trade mark applications.
d) Word or figurative element trade mark searches.
e) Databank service
    – Searches for scientific and technical information, patent and commercial protection rights, market intelligence.
    – Technical analysis and evaluation.
    – Monthly monitoring of databanks for new developments.
    – Instruction, training and exchange of information seminars.

RATINGEN

Zentralstelle für Textildokumentation und -information
Schloß Cromford
Cromforder Allee 22
D-4030 Ratingen
Tel. (02102) 2 70 51
Telex: 8 858 374 vidi d

Open: Mo – Fr 9.00 – 12.00 and 14.00 – 16.00

Patent documents
OS, AS, PS in selected classes only; DE-EP, DE-WO

Literature
Special to textile technology

STUTTGART

Landesgewerbeamt Baden-Württemberg
Kienestraße 18
D-7000 Stuttgart 1

Tel. (0711) 123 0
Telex: 0723931 lga
Telecopier: (0711) 2020-2560

Open: Tu, Th, Fr 10.00 – 16.00, We 10.00 – 19.00, Sa 10.00 – 12.30

Patent documents
OS 1 400 000 – (1. 10. 68) – ; *AS* 100,001 (1957) – ; *PS* 1877 – ; *Gbm* 1964 – ;
DE-EP; DE-WO
AT 1926 – ; CH 1888 – , EP 1980 – , FR 1968 – , WO 1980 –
Patent abstracts of Japan
Official gazettes: AT 1902 – 15, 1956 – , CH 1889 – , DE 1877 – ,
　　　　　　　　　EP 1978 – , US 1920 – , WO 1978 –

Literature
BlPMZ 1894 – ; Mitt 1954 – ; GRUR 1948 – ; GRURInt 1980 – ;
Official Journal of the EPO 1978 –
Also many books and periodicals.

Standards and Guidelines: DIN, VDE, VDI

INPADOC: PAS, PIS, IPG, NDB (DE, AT, CH, EP, Gbm)

Online access to INPADOC, CLAIMS, Derwent

Free consultation with a patent agent every Wednesday 10.30 – 12.00

Patent databank searches are made on behalf of small and medium sized firms
and independent inventors.

Wila-Verlag Wilhelm Lampl KG   **DE**

Landsberger Straße 191a
D-8000 Munich 21
Tel: (089) 57 951
Telex: (05) 212 943

Abstract journals published by Wila-Verlag

The following abstract journals cover all published German patent applications *(Offenlegungsschriften, OS)*, accepted patents *(Patentschriften, PS)* and registered utility models *(Gebrauchsmuster, Gbm)* as well as European patent applications *(Europäischen Patentanmeldungen, EPZ)* and European patents *(Europäischen Patentschriften, EPS)*. The abstracts include all bibliographic data (IPC, patent number, filing date, publication date, priority, title, inventor, applicant, etc.), the most significant (first) patent claim and the most important drawing or formula.

Auszüge aus den Offenlegungsschriften   Oct. 1968 –   weekly

Part I | Raw materials, chemistry, mining, building,
(IPC: A61 K and L, B01 B-0, BO3, BO7 B, BO8, B21, B22, B28, B32, C complete, D complete, E complete except E05.)

Part II | Electrotechniques, physics, fine mechanics, optics, acoustics.
(IPC: A61 B-J, A61 M-N, A62, A63, B01 K-L, B06, B41, B43, B44, B61 L, G complete, H complete.

Part III | Nutrition, agriculture, machines and vehicles, other processing industries.
(IPC: A complete except A61-63, B02, B04, B05, B07 C, B23-27, B29-31, B42, B60-68 except B61 L, E05, F complete.)

Subscription: each part DM 130, monthly; two parts 10% reduction; three parts (complete issue) 20% reduction.

Card file: The abstracts for each IPC class can be supplied weekly on file cards (DIN A6). Subscription rates depend on the extent of the class.

Auszüge aus den Patentschriften   Sept. 1955 –   weekly

Subscription: Edition A (printed on both sides of the page) DM 92, monthly.
Edition B (printed on one side of the page) DM 112, monthly.

Auszüge aus den Gebrauchsmustern   Oct. 1964 –   weekly

Subscription: Edition A (printed on both sides of the page) DM 92, monthly.
Edition B (printed on one side of the page) DM 112, monthly.

Auszüge aus den Europäischen Patentanmeldungen   Jan. 1985 –   weekly

Parts I, II and III – subject coverage of each part is the same as *Auszüge aus den Offenlegungsschriften* above.

Subscription: each part DM 92, monthly; two parts 10% reduction; three parts (complete issue) 20% reduction.

Auszüge aus den Europäischen Patentschriften   Jan. 1980 –   weekly

Subscription: printed on both sides of the page, DM 96, monthly.

DENMARK   DK

PATENT OFFICE

PATENTDIREKTORATET
Nyropsgade 45
DK-1602 København V
Tel: (01) 128440
Telex: DPO DK 16046

I General information

The Danish Patent Office is an office under the Ministry of Industry. It comprises three divisions: The Industrial Property Division, the Information and Documentation Division and the Administrative Division. It is an examining patent office and is responsible for the grant of patents in Denmark, including the Faroe Islands and Greenland. Classified and numerical collections of patent documents are available to the public. Searches may be requested from the Service Section, a section within the Information and Documentation Division.

Staff (total):   239
Patent examiners: 72
Library staff: 21

II Patent documents

Almindelig tilgængelig patentansøgning (A)   1968 –

Danish patent application made available to the public 18 months after the date of filing or, where priority is claimed, the date of priority. The applications are published in accordance with the Consolidate Patents Act, 1986, section 22 (2) and (3) or corresponding provisions in earlier Patents Acts. Applications from Danish residents are searched, but not examined before publication. Other applications are normally neither searched nor examined before publication. The applications are not printed, but photocopies are supplied on request.

Fremlæggelsesskrift (B)   1968 –

Danish patent application laid open to public inspection to allow opposition by third parties.

(19) DANMARK

(12) FREMLÆGGELSESSKRIFT (11) 149509 B

DIREKTORATET FOR
PATENT- OG VAREMÆRKEVÆSENET

(21) Patentansøgning nr.: 0445/82

(22) Indleveringsdag: 02 feb 1982

(41) Alm. tilgængelig: 04 aug 1982

(44) Fremlagt: 07 jul 1986

(86) International ansøgning nr.: –

(30) Prioritet: 03 feb 1981 DE 3103544

(51) Int.Cl.⁴: D 06 F 65/02

(71) Ansøger: WILH. °CORDES GMBH & CO. MASCHINENFABRIK; Oelde, DE.

(72) Opfinder: Werner °Schulze; DE, Franz W. °Walterscheid; DE.

(74) Fuldmægtig: Patentbureauet Magnus Jensens Eftf.

(54) Strygemaskine

(57) Sammendrag:

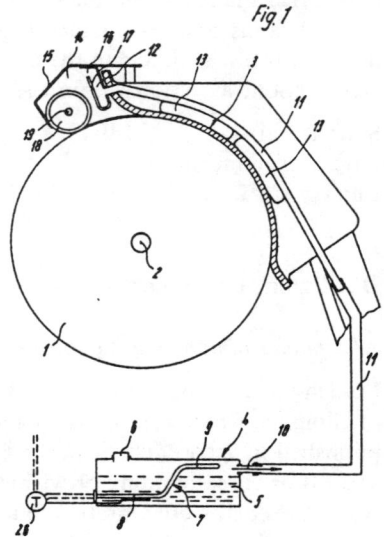

Fig.1

445-82

En strygemaskine er på indføringssiden for det, som skal
stryges, udrustet med et dæmpningsrum (14), som er udformet
åben i retning mod en strygerulle (1) og på de andre sider er
begrænset ved hjælp af et hus (15). Dampsperring kan på indføringssiden være tilvejebragt ved hjælp af en i huset dreje-
lig lejret og med radialt spillerum anordnet rulle (18). I
dæmpningsrummet (14) er der anbragt et fordelerrør (12), som
i det øvre område er i besiddelse af en udstrømningsåbning
(16) for dampen, som er underkastet tilførsel til fordelerrø-
ret (12) via en dæmpledning (11) og er frembragt i et fordampningskammer (5). Dette fordampningskammer (5) er udrustet
med forkrøbbet udformede, stavformede varmeelementer (7),
hvoraf en gren (8) ligger i den påfyldte vandmængde, og en
gren (9) ligger uden for denne vandmængde.

DK 149509 B

62

The applications are published in accordance with the Consolidate Patents Act, 1986, section 21(1) and (3) or corresponding provisions in earlier Patent Acts. They are examined applications which are published in multiple copies (printed).

Patentskrift (C)   1968 –

Danish patent specification.

The specifications are published in accordance with the Consolidate Patents Act, 1986, section 26 (1) or corresponding provisions in earlier Patents Acts. They are specifications of granted patents based on examined applications and are published only when the text differs from that of the corresponding applications laid open to public inspection. If so, they are published in multiple copies (printed). Otherwise "Patent granted" is stamped on the applications laid open to public inspection.

Patent   1895 –

Danish patent specification

These are specifications of granted patents based on examined applications and published in accordance with the Patents Act 1894 as subsequently amended. The latest in the sequence of printed specifications was published on 12 June 1978; a few more may still be published.

Since January, 1973, all bibliographic data are presented on the front page of the patent documents in accordance with WIPO Standard ST. 9.

Since 1975 patent applications laid open to public inspection (B documents) and patent specifications (C documents) are available on 8-up aperture cards.

III Official gazette

Dansk Patenttidende (Danish patent gazette) 1894 –   weekly

Table of contents

– *Nye patentansøgninger*	New patent applications
– *Almindeligt tilgængelige patentansøgninger*	Patent applications made available to the public
– *Fremlagte patentansøgninger*	Patent applications laid open to public inspection
– *Udstedte patenter*	Patents granted
– *Tilbagetagne, afslåede og henlagte patentansøgninger, hvis akter er almindeligt tilgængelige*	Patent applications withdrawn, refused or dismissed, their documents being available to the public
– *Rettigheder ifølge patentansøgninger og patenter genoprettet i henhold til patentlovens § 72*	Rights conferred by patent applications and patents re-established under section 72 of the Consolidate Patents Act, 1986

– Bortfaldne patenter	Patents lapsed
– Udløbne patenter	Patents expired
– Navneregister	Name index
– Internationale patentansøgninger, hvori Danmark er designeret	International patent applications designating Denmark
– Andre meddelelser	Other announcements: Occasional announcements of essential decisions made by the Patent Board of Appeal and other matters concerning the filing and processing of patent applications

Tillæg til Dansk Patenttidende (Supplement to the Danish patent gazette)

Issued weekly since the 4th September, 1968

The supplement contains abstracts and, where appropriate, drawings of patent applications made available to the public. The supplement is divided into sections A to H of the International Patent Classification, section C being further subdivided into two subsections.

Ugeliste (Weekly list of patent applications filed)

Numerical list of patent applications filed within a certain week and of international patent applications designating Denmark.

Register over danske patenter (Index of Danish patents)   1894 –   annual

Table of contents

– Name index of patents granted
– Classified index of patents granted
– Numerical list of patents granted indicating their classification

IV Source of supply and prices

Publications and photocopies are obtainable from:

PATENTDIREKTORATET
Nyropsgade 45
DK-1602 København V

                                                                          DKK

Fremlæggelsesskrifter (Patent applications laid open to
public inspection) including patent specifications with differing text)
– annual subscription   3000. –
– by subscription to classes or subclasses per copy   5. –
– single copy   10. –

Dansk Patenttidende (Danish patent gazette)
- annual subscription (Danish subscribers)   500. –
- annual subscription (foreign subscribers)   600. –
- single copy   20. –

Tillæg til Dansk Patenttidende (Supplement to the Danish patent gazette)
- annual subscription to:
    section A(Human necessities)   600. –
    section B (Performing operations; Transporting)   600. –
    section C07 and C12 (Organic chemistry; Biotechnology)   600. –
    section C except C07 and C12 6 Chemistry; Metallurgy)   300. –
    section D (Textiles; Paper)   150, –
    section E (Fixed constructions)   300. –
    section F (Mechanical engineering; Lighting; Heating; Weapons; Blasting)   300. –
    section G (Physics)   300. –
    section H (Electricity)   300. –
    all sections   3000. –

Register over Danske Patenter (Index of Danish patents)   300. –

Ugeliste (Weekly list)   10. –

Photocopies of other documents per page   2. –

The prices of publications and photocopies are inclusive of postage.

Orders for photocopies are normally executed within three days; library users may avail themselves of a self-service system at the same price.

The Patent Office does not require any deposit or prepayment before executing orders for photocopies. However, if desired, it is possible to open an account with the Office, on which an amount corresponding to at least the estimated expenditure over three months must be deposited.

The said deposit may also be used to pay for single copies of printed patent documents and for services obtained from the Service Section, but not for subscriptions or fees. The account holder must ensure that the price of ordered copies, etc. does not exceed the remaining deposit. An invoice is enclosed with every supply of copies.

V   Register of legal status

The patent office keeps a record of patent applications filed and a register of patents granted.

The record of applications is open to the public for inspection or for telephone enquiry free of charge.

The public may obtain certified copies of the register of patents at a price of DKK 50. – per entry.

From 1987 the record of applications and the register of patents will be accessible online. The price for this service has not yet been fixed.

VI Public services

a) *The Library*
   The reading room of the Danish Patent Office is open: –
      Monday 9.00 – 14.00 and 17.00 – 19.00
      Tuesday to Friday 9.00 – 14.00

There is no entrance fee: photocopy costs DKK 2. per page.

Librarians with thorough knowledge of the various classification systems will assist the library users. In special cases examiners may be called to assist.

The following collections are directly accessible in the reading room:

- Classified and numerically ordered sets of printed Danish patent documents.
- On-line access to data bank of bibliographical data of all Danish patent applications filed on and after the 1st January, 1982, regardless of their status, and of Danish patent applications filed earlier if pending on the 1st January, 1984.
- Card indexes of Danish patent applications not accessible on-line, viz. index according to applicants' names ordered alphabetically and index according to classification (from 1968 to 1971 according to the German Patent Classification, from 1972 and onwards according to the International Patent Classification).
- IPC, German, Dutch and US classification manuals.
- Danish and various foreign patent office journals and gazettes.

The classified files of the Danish Patent Office comprise patent documents from the following countries or organizations:

Classified according to German classification:

Denmark
Norway   up to 1971
Sweden
Federal Republic of Germany

Classified according to Dutch classification:

France (old law publications)   up to 1973
United Kingdom (abridgments of old law publications)   up to no distinct
                                                                   year

Classified according to IPC:

Denmark	
Norway	
Sweden	since 1972
Federal Republic of Germany	
Finland	since 1944
France (new law publications)	since 1969
United Kingdom (front pages of 1977 Act publications)	since 1978
EPO	since 1978
WIPO (PCT)	since 1978

Classified according to US classification:

USA	since 1909

At present there is no public access to the United Kingdom abridgments of old law publications referred to above.

All the classified files referred to may be brought to the public reading room for examination.

Other patent documents are only available in numerical sets.

On request books and periodicals on industrial property law may be brought to the public reading room for studying.

Patent documents in the library of the Danish Patent Office

AT	Austria	specns	1899 –
		gazette	1899 –
AU	Australia	specns	1937 –
		gazette	1904
BE	Belgium	specns	1950 –
		gazette	1914 –
BG	Bulgaria	specns	1965 –
		gazette	1967 –
CA	Canada	abstracts	1922 –
		gazette	1912 –
CH	Switzerland	specns	1888 –
		gazette	1889 –
CS	Czechoslovakia	specns	1947 –
		gazette	1919 –
DD	German Democratic Republic	specns	1951 –
		gazette	1960 –
DE	Germany, Federal Republic of	specns	1877 –
		gazette	1882 –
DK	Denmark	specns	1895 –
		gazette	1894 –

EP	European Patent Office	specns	1978 –
		gazette	1978 –
FI	Finland	specns	1944 –
		gazette	1889 –
FR	France	specns	1900 –
		gazette	1890 –
GB	United Kingdom	specns	1893 –
		gazette	1893 –
HU	Hungary	specns	1940 –
		gazette	1896 –
IT	Italy	gazette	1907 –
JP	Japan	specns (A)	1951 –
		Pat. Abstr. of of JP	1976 –
NL	Netherlands	specns	1912 –
		gazette	1912 –
NO	Norway	specns	1886 –
		gazette	1886 –
NZ	New Zealand	gazette	1912 –
PL	Poland	specns	1941 –
		gazette	1924 –
SE	Sweden	specns	1885 –
		gazette	1902 –
SU	Soviet Union	specns	1968 –
		gazette	1958 –
US	United States of America	specns	1895 –
		plant specns	1931 –
		gazette	1894 –
WO	WIPO-PCT	specns	1978 –
		gazette	1978 –
YU	Yugoslavia	specns	1961 –

b) The Service Section

The Service Section offers primarily the services listed below, but other services may be rendered by the Section if need should arise.

The investigations necessary for answering the questions put to the Service Section are made by members of the technical staff of the Patent Office (patent examiners) when technical problems are involved. Otherwise they are made by the clerical staff of the Service Section.

When the investigation concerning a particular case is completed a report is drawn up and forwarded to the inquirer. The report will specify the retrieved documents as well as the scope of the search as to countries, classes and periods of time.

Types of investigations

a) Novelty searches based on a short description of an invention
The report will refer to disclosures in patent specifications and applications considered relevant.
The result of the search will make it possible for the client to estimate whether or not he should file a patent application.
It should be emphasized that the report will only concern the technical contents of the retrieved documents and will not comment on the novelty of patentability of the invention searched. The evaluation of that is left to the client or his advisor.

b) Novelty searches for the purpose of evaluating whether it is reasonable to file an opposition in a foreign country or to contest the validity of a patent granted in a foreign country.

c) Investigations for industrial firms intending to put a new product on the market and wanting to know whether such marketing will infringe any existing patent rights.

d) Investigations useful for the development of products
The report will give information on recent technology in the field in question as indicated by the patent literature.

e) Searches by names
The Service Section may advise whether a certain company or person is the holder of a patent in certain countries (at present 52 countries, i.e. most countries which have acceded to the Paris Convention) or has a pending patent application in such countries. Furthermore, the Service Section may advise whether a certain company or person is the holder of patents or has applications pending within a certain technical field.

f) Investigations for the purpose of ascertaining whether a specified invention is patented or is the subject of a pending application in Denmark.

g) Patent family investigations
Patents or patent applications in various countries concerning one and the same invention are said to make up a patent family. On request the Service section will search for patents and pending applications belonging to a certain patent family.

h) Current awareness service
The Service Section undertakes many types of current awareness service concerning Danish as well as foreign patents or patent applications.
For example, the Service Section may give notice
  – whenever a patent or a patent application relating to a certain technical field is published in a certain country or in the name of a certain company or firm
  – whenever there is a change of status with respect to a certain Danish patent application.

– if a certain European patent application proceeds to grant; in that case the notice will be given in good time before the expiry of the opposition period.

i) Technical assistance
It is possible to "hire" an examiner for a few hours to assist clients with searches in the classified files of the Patent Office in the public reading room.

According to an agreement between the Nordic Patent Office and INPADOC, the latter provides the Danish Patent Office with the following documentation aids:

PFS   Patent Family Service
PAS   Patent Applicant Service
PIS   Patent Inventor Service
PCS   Patent Classification Service
NDB   Numerical Data Base
INL   INPADOC Numerical List
IPG   INPADOC Patent Gazette

With these aids searches by name (item e) may be extended to include inventors, and patent family investigations (item g) may be extended to include applications for which priority has not been claimed.

The fees are normally calculated according to the time involved. Present rates: DKK 150 – 300 per hour.
For a few services, however, a flat rate is charged, e.g. DKK 250 for each patent family search in the INPADOC material.

Routine inquiries may be answered within one or two weeks, while novelty searches and the like may take two to four weeks.

c) Data banks available to the public

Data base comprising Danish patent applications and patents

At the end of 1986 new computer equipment will be installed in the Danish Patent Office. In the course of 1987 the record of patent applications and the register of patents will be converted to a database to which the public will have access. The database will include the bibliographic data published in the official gazette as well as other data relevant to patent procedure.

Other data bases

The Service Section offers technical searches in and the establishment of search profiles for a number of data bases at various hosts to which the Danish Patent Office has access.

Examples of database hosts are: SDC, Lockheed Retrieval System, ESA-IRS, BRS, Télésystèmes-Questel and Pergamon InfoLine.

Furthermore, the Service Section offers bibliographical searches in the Derwent databases WPI and WPIL, in the INPADOC databases and in the EPO patent database.

VII Provincial libraries

Libraries holding the Danish official gazette and Danish patent documents.

	Official gazette	Suppl. to official gazette (abstracts of patent applications)	Patent specifications *)
HERNING: Herning Centralbibliotek Brændgårdvej 2 7400 Herning Tel. (07) 12 18 11	1981 –		1981 –
HORSENS: Horsens Bibliotek Vitus Berings Park 8700 Horsens Tel. (05) 61 43 44	1938 –	1978 –	1938 –
ODENSE: Odense Tekniske Bibliotek Niels Bohrs Alle 11 5230 Odense M Tel. (09) 13 08 27	1937 –	1978 –	1937
RØNNE: Bornholms Centralbibliotek Pingels Alle 1 3700 Rønne Tel. (03) 95 07 04	1953 –	1978 –	1953 –
TÓRSHAVN: Føroya Landsbókasayn J.C. Svabosgøtu 16 3800 Tórshavn Tel. (042) 1 16 26	1956 –		1956 –

*) From 1975 applications laid open to public inspection (fremlæggelsesskrifter)

VEJLE:
Biblioteket for Vejle By og   1937 –   1937 –
Amt
Vestre Engvej 55
7100 Vejle
Tel. (05) 85 32 00

AALBORG:
Biblioteket på Grønlands
Torv   1937 –   1978 –   1937 –
Grønlands Torv
9210 Aalborg SØ.
Tel. (08) 14 15 12

ÅRHUS:
Århus Kommunes Biblioteker   1930 –   1978 –   1930 –
Mølleparken
8000 Århus C.
Tel. (06) 13 66 22

EUROPEAN PATENT OFFICE
Erhardtstraße 27
D-8000 Munich 2
Tel.: (089) 23 99-0
Telex: 52 36 56 epmu d

Branch at the Hague (DG1)
Patentlaan 2
Postbus 5818
2280 HV Rijswijk ZH
Netherlands
Tel.: (070) 40-2040
Telex: 31 651 epo nl

Sub-office in Berlin
Gitschinerstraße 103
D-1000 Berlin 61
Tel.: (030) 25 89 01

I   General information

The European Patent Organisation came into being on 19 October 1977, shortly after the entry into force of the European Patent Convention on 7 October 1977. Its headquarters are in Munich, with a branch at the Hague and a sub-office in Berlin.

The Organisation is composed of a supervisory body, the Administrative Council, which represents the Contracting States to the Convention, and the European Patent Office (EPO) which has administrative and financial autonomy. Its task is to grant European Patents which have the effect of, and are subject to the same conditions as, national patents in the Contracting States.

An applicant, by designating any or all of these Contracting States in a patent application can obtain the same patent protection as available under the relevant national patent law. National patent law in the Contracting States has been harmonised with the provisions of the European Patent Convention and a granted European patent is effectively a "bundle" of national patents in the designated Contracting States.

Europäisches Patentamt

European Patent Office

Office européen des brevets

(11) Publication number: **0 217 507**
A1

(12) **EUROPEAN PATENT APPLICATION**

(21) Application number: 86305913.5

(22) Date of filing: 31.07.86

(51) Int. Cl.⁴: **B64G 1/14 , B64G 1/62**

(30) Priority: 05.09.85 US 772802

(43) Date of publication of application:
08.04.87 Bulletin 87/15

(84) Designated Contracting States:
DE FR GB IT

(71) Applicant: GRUMMAN AEROSPACE
CORPORATION
South Oyster Bay Road
Bethpage, NY 11714(US)

(72) Inventor: Corbett, Marshall J.
15 Elberta Drive
East Northport New York 11731(US)

(74) Representative: Andrews, Robert Leonard et
al
HASELTINE LAKE & Co. Hazlitt House, 28
Southampton Buildings Chancery Lane
London WC2A 1AT(GB)

(54) Manned entry vehicle system.

(57) A manned space entry vehicle having pivotally
mounted wings (36) which remain stored during
space travel and become deployed to a forward-
swept configuration after entry in the atmosphere. A
radar antenna (FIG. 3) is mounted in the forward
portion of the wing so that it is protected from
excessive heat when the wing remains in a stored
position during early reentry into the atmosphere.
When the wings are deployed, the radar antenna
maintains an operative position. The mean aero-
dynamic center of the deployed wings is transverse-
ly aligned with the center of gravity of the vehicle so
as to increase the stability of the vehicle. The basic
vehicle shape is intended to produce the least reflec-
tive radar cross section as it passes over threatening
areas in a nose down attitude.

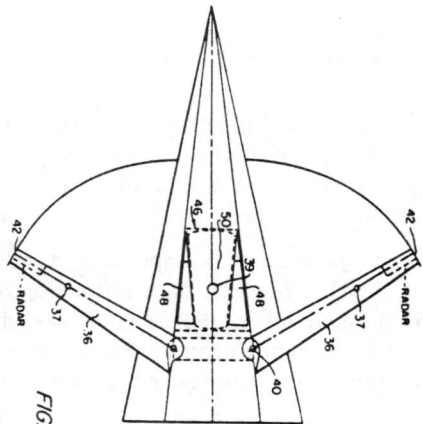

FIG.3

EP 0 217 507 A1

Instead of a separate procedure before the national office of each State in which protection for the invention is sought, it is now possible to file one application with the EPO. A single grant procedure, carried out in either English, French or German replaces national procedures.

From 1 October 1986 the following thirteen Contracting States are members of the European Patent Organisation:
Austria, Belgium, Federal Republic of Germany, France, Greece, Italy, Liechtenstein, Luxembourg, Netherlands, Spain, Sweden, Switzerland, United Kingdom.

The European Patent Office has five Directorates-General,
DG1 – Search, DG2 – Examination/Opposition, DG3 – Appeals, DG4 – Administration, DG5 – Legal/International Affairs.
DG1 is located at the Hague and DGs 2–5 are in Munich.

DG1, together with the sub-office in Berlin, deals with the first phase of the grant procedure, i.e. the search and publication of the application and its search report. This branch is also a Receiving Office and International Search Authority under the Patent Cooperation Treaty.

DG1 has taken over the work of the former International Patent Institute (IIB) and carries out searches on Dutch, French and Swiss applications, and shortly also Belgian. Applicants, irrespective of nationality or domicile, can request DG1 to prepare search reports on the state of the art either through the European patent grant procedure or by requesting special searches in the documentation.

Staff at the beginning of 1986
Munich (total): 932 (substantive examiners 432)
The Hague (total): 945 (search examiners 530)
Berlin (total): 129 (search examiners 61)

II Patent documents

– European patent application (A publication) containing the description, claims, abstract and any drawings as filed as well as the European Search Report which may be published separately if not available in time.
– Specification of the granted European patent (B publication) containing the description, claims in all three official languages, and drawings. There is the possibility of publication of an amended specification following opposition proceedings.

Note that applications reaching the EPO via the PCT route i.e. Euro-PCT applications, are available as PCT documents and are not republished as A publications by the EPO if they are in English, French or German.

Applicants for European patents are invited to file, in addition to the normal paper form, also on diskette or in machine readable OCR typestyle. Details of the DATIMTEX (data, images and text carriers) system are obtainable from DG1 at the Hague, ext. 2282 or 3089, or from DG4 at Munich, ext. 4111.

III Official gazette (other publications are listed in Section IV)

Official Journal of the European Patent Office   1978 —   monthly

This is a trilingual journal (German, English, French) containing notices from the President and general information related to the European Patent Convention. The contents are grouped under the following headings:
Administrative Council; Decisions of the Boards of Appeal; Information from the EPO; Information from the Contracting States; International Treaties; Fees; Vacancies; Advertising.

European Patent Bulletin   1978 —   weekly

This journal is also trilingual and contains bibliographic data on published European patent applications and granted patents, including selected information on their procedural status.

Contents list

Part I   Published applications
    I.1   Full bibliographic data arranged by IPC
      2   International applications arranged by PCT publication number.
      3i   Arranged by EP publication number with IPC
      ii   Arranged by EP application number with publication number
      4   Arranged by name of applicant
      5   Arranged by designated Contracting State with EP publication number and IPC
    I.6 – 12   Stages of progress of applications listed under EP publication number
            6 Search report
            7 Examination
            8 Refusals, withdrawals, etc
            9 Suspensions, interruptions, etc
            10 Conversions
            11 Licences
            12 Changes/rectifications

Part II   Granted European patents

II.1   Full bibliographic data arranged by IPC
    2i   Arranged by EP publication number with IPC
    ii   Arranged by EP application number with publication number

3   Arranged by name of proprietor

4   Arranged by designated Contracting State with EP publication number and IPC

II.5 – 12 Stages of progress of patents listed under EP publication number

          5   Lapse in Contracting State

         6 – 7 Opposition, revocation, maintenance

          8   Re-establishment of rights

          9   Suspension, resumption, interruption

         12   Changes/rectifications

Name Index to the European Bulletin   1978 –   annual

Part I   Published applications

Lists, in alphabetical order of names of applicants (in bold type), the applications published in the *European Patent Bulletin* with the number of the relevant *Bulletin,* the name of the inventor (in normal type), the title of the invention and the publication number of the specification. Numbers marked with an asterisk are Euro-PCT applications for which the WO number is also given.

Part II   Granted patents

Lists similarly names of proprietors and inventors of granted patents.

IV Sources of supply and prices – from 3 January 1985

		DM
1. *Official Journal of the European Patent Office*		
	Subscription, in Europe	125
	outside Europe	175
2. *European Patent Bulletin*		
	Subscription, in Europe	430
	outside Europe	640
3. *Annual name index* to the *European Patent Bulletin*		
	in Europe	50
	outside Europe	60
4. European patent applications (A)		
	Single sales	6.50
Available singly or by IPC group. Special EPO order forms may be used.		
5. European patent specifications (B)		
	Single sales	13
Also available singly or by IPC group.		
6. European patent applications and European patent specifications on aperture card		
	Single sales	4.5
Also available singly or by IPC group.		
Supplied about 4 to 6 weeks after publication of the relevant printed versions.		

Publications listed at 4, 5 and 6 may only be supplied on subscription provided the subscriber has a deposit account with the European Patent Office (Cash and Accounts Department). Details will be supplied on request.

Other publications of the European Patent Office

DM

Guidelines for examination in the European Patent Office
Current Office practice as regards search, examination and opposition procedures for the grant of European patents. Published in English, French and German and updated regularly.
In loose-leaf plastic ring binder (19.5 x 25 cm).

Single sales 55

European Patent Convention   3rd ed. Jan. 1985
Text of the Convention, the Implementing Regulations, the Protocol on Centralisation, the Protocol on Recognition and the Rules relating to Fees.
Tri-lingual, cloth bound (17 x 24).

Single sales 32

Directory of Professional Representatives   6th ed. Jan. 1986
Single sales 17

Cross-reference index
For PCT applications in accordance with Article 158, para 1 of the EPC, arranged by European publication number (List A) and PCT publication number (List B).
Copies of EDP print-outs in A4 format, per page

1.30

The above publications are obtainable from:

European Patent Office
Department 4.5.2 (Distribution)
Erhardtstraße 27
D-8000 Munich 2

Free publications of the European Patent Office

* *How to get a European Patent*
  Guide for applicants describing application and grant procedure. Published in English, French, German, Italian and Japanese.

* *National law relating to the EPC*
  Synopsis of the regulations and requirements in the Contracting States concerning European patent applications and patents. Published in English, French and German.

* *Protecting inventions in Europe*
  Information leaflet giving general introduction to the European patent

system. Published in English, French, German, Italian, Dutch, Spanish and Swedish.

* *Development co-operation*
Information leaflet providing an introduction to the activities of the European Patent Organization in the field of technical assistance for developing countries. Published in English, French and Spanish.

* *The European Patent Office*
General introduction to the European Patent Office building in Munich, the branch at the Hague and the sub-office in Berlin. Trilingual edition.

* *The European Patent Office DG1*
Describes the origins and functions of the branch at the Hague with information on official search procedures and search and information services available to industry. In English.

These free publications, marked *, are obtainable from the European Patent Office at Munich, Department 4.5.1 (Public Information).

Video films in the three official languages of the Office on the European grant procedure are available on loan from Department 5.2.1 following written request which should specify the language and video format required.

Outside publishers

The following publications contain official EPO information and are obtainable from the outside publishers named.

Auszüge aus den Europäischen Patentanmeldungen

Abstracts of European patent applications published weekly in the language of origin from 1985 in three parts by Wila-Verlag. For details see under Wila-Verlag.

EPO Applied Technology Series

The EPO has prepared a series of patent surveys relating to selected areas of technology. The current list includes optical fibres, dynamic semiconductor RAM structures, solid state video cameras, silicon nitride and silicon carbide ceramics, methods of abating residual formaldehyde in industrial resins, vaccines for viral hepatitis, nickel and cobalt extraction using organic compounds, microprocessors, industrial robots, inorganic fibres and composite materials, reverse osmosis. These surveys are published by Pergamon Press from whom further details may be obtained.

Pergamon UK	or	**Pergamon USA**
Headington Hill Hall		**Fairview Park**
Oxford OX3 OBW		**Elmsford**
England		**NY 10523**
		USA

Decisions of the Boards of Appeal of the European Patent Office

The collection of published decisions from the *Official Journal of the EPO* of the Legal Board of Appeal, Technical Boards of Appeal and Disciplinary Board of Appeal is available from Carl Heymanns Verlag. The decisions comprise the German, English and French text as published. Details may be obtained from:

Carl Heymanns Verlag
Steinsdorfstraße 10
D-8000 Munich 22

V Register of legal status

As required by the EPC an extensive list of register entries concerning published European applications and patents is open to public inspection. The register is available during normal office hours at the information desks in Munich and The Hague. Telephone enquiries may be made on 089- 2399-4538 at Munich or 070-40-3259 at The Hague.

Direct access by terminal to the register is possible on payment of a subscription fee. Full information on this service is available from Directorate 4.5 at Munich.

It is proposed to make access to the register available to the national industrial property offices of the Contracting States.

VI Public services

Searching

As part of the normal equipment in connection with the preparation of state of the art searches and search reports the branch at The Hague has a defined collection of patent and non-patent literature. The following information systems have been organized for online use by EPO examiners and may also be available for public access through the public information services of the patent offices of EPO Member States:

- The FAMI system, which enables patent documents classified in the search documentation to be identified as belonging to the same patent family, and indicating the classification symbols allotted by the EPO.
- The INVE system, indicating which documents are classified under a particular classification symbol.
- The ECLA system, comprising the text of the classification, including its symbols, used by the EPO.

The *standard search* service of DG1 is principally available to applicants whereby a patent search is performed in accordance with the guidelines for European patent applications, but not as part of a European patent application as such.

In this way the applicant for a standard search may acquire the state of the art with respect to a particular invention provided a patent application has been filed in any country. The fee for a standard search is higher than the fee for a European search forming part of a European application, but may be set against the search costs of a subsequent European filing.

Special searches in the EPO search documentation carried out by DG1 personnel are also available on a restricted basis. These can be ordered to locate patenting patterns in certain technical fields, to draw up inventories of the contents of parts of the classified documentation, to carry out preliminary searches prior to an industrial research programme, etc. Fees are usually charged on a time-cost basis and requests for estimates should be sent to DG1 at The Hague.

Visitors' programmes

The European Patent Office is willing to arrange information visits to Munich and The Hague for interested groups of patent specialists and has an active policy of co-operation activities for the benefit of developing countries. Such visits may be arranged by writing, mentioning specific interests, to Department 521, Press and Public Relations, at Munich or the Press and Public Relations Department at The Hague.

Library

The library at Munich is principally for staff use but some of the material is available to the public. Information requests should initially be addressed in every case to the library. There is a numerical collection of European patent specifications and a numerical and classified collection of PCT specifications as well as literature on technology, natural sciences and industrial property law.

Open   Monday to Thursday 8.00 – 16.45 and Friday 8.00 – 15.30

The library at The Hague is that of the Netherlands patent office.
Open   Monday to Friday 9.00 – 12.00 and 1.30 – 17.00

PATENT OFFICE

REGISTRO DE LA PROPIEDAD INDUSTRIAL
Panama, 1
28036 Madrid
Tel. 4 58 22 00

I   General information

The Registro de la Propiedad Industrial (RPI) is an autonomous body forming part of the state administration under the Ministerio de Industria y Energia (Ministry of Industry and Energy) and is responsible for the administration of the various forms of industrial property in Spain.

Staff: 325

II   Patent documents

Patents and utility models are not printed but are laid open to public inspection at the RPI after grant, photocopies or microfilm copies of the original specifications are available on request.

III   Official gazette and other publications

Boletin Oficial de la Propiedad Industrial (BOPI)   1886 –   twice monthly

It is published in three parts:

I   Trade marks and other distinctive signs
II   Patents and utility models
III Industrial and artistic models and designs

The following are published in Volume II:

| Sumario | Table of contents |

Sumario

*I. Patents de invencion, de introduc-
ción y certificados de adición*

— *Suspensos*
— *Resoluciones:*
  ○ *Concesiones (número, titular
  y domicilio, enunciado de la
  invención, prioridades,
  Clasificación Internacional,
  fecha de solicitud y fecha de
  concesión)*
  ○ *Anulaciones, caducidades,
  denegaciones*
  ○ *Ofrecimientos de licencia*
  ○ *Decisiones adoptadas en
  virtud de recursos o por
  sentencias de Tribunales*

2. Modelos de Utilidad
— *Solicitudes:*
  ○ *Datos de identificación
  reivindicaciones y disenos*
— *Suspensos*
— *Resoluciones*
  ○ *Concesiones (figuran los
  mismos datos que en
  patentes, asi como tambien
  la fecha de publicación de la
  solicitud en el BOPI)*
  ○ *Anulaciones, caducidades,
  denegaciones*
  ○ *Ofrecimientos de Licencia*
  ○ *Decisiones adoptadas en
  virtud de recursos o por
  sentencias de Tribunales*

Table of contents

I. Patents of invention, patents of
importation and certificates of addition
— Patents pending
— Decisions:
  ○ Patents granted (number, name
  and address of patentee,
  announcement of invention,
  priorities, IPC, date of
  application and date granted)
  ○ Cancellations, forfeitures and
  refusals
  ○ Licences of rights
  ○ Decisions adopted by virtue
  of appeals or court judgments

2. Utility models
— Applications
  ○ Information on identification
  claims and designs
— Utility models pending
— Decisions:
  ○ Utility models granted (Including the same information
  as for patents, plus the date
  of publication of the
  application in the BOPI)
  ○ Cancellations, forfeitures and
  refusals
  ○ Licences of right
  ○ Decisions adopted by virtue
  of appeals or court
  judgments.

Información Tecnológica de Patentes   1981 —   twice monthly

Contains abstracts of granted patents.

Indices anuales   1965 —
  Part I   Patents of invention, patents of importation and certificates of ad-
      dition.
  Part II   Utility models.
  Part III   Industrial and artistic models and designs.

The index includes the following bibliographic data in name, numerical and classified order: −

− number of patent or utility model
− name of proprietor
− IPC
− title of the invention
− date of publication of issue in BOPI
− date of publication of application in BOPI (utility model)

Clasificación internacional de Patentes, cuarta edición. 1984.

The 4th edition of the IPC in Spanish is published by the RPI jointly with the World Intellectual Property Organization.

IV Source of supply and prices

Prices are published in the *Boletin Oficial del Estado*, the most recent on 21 July 1984.

Publications and photocopies may be requested by mail or telex from: −

Registro de la Propiedad Industrial
Oficina de Difusión
Panama, 1
28036 Madrid

Telex: 47020 RPIE

V Register of legal status

The RPI has developed SITADEX, a database containing more than 500,000 records on the legal status of all industrial property rights registered in Spain. This database may be used by the public at the Office free of charge.

VI Public services

The RPI has a reading room open to the public and providing: −

− information on Spanish patents; although Spanish patents are not printed photocopies of the specifications are available on request.
− information on foreign countries with patent documents available for the following authorities: −

AR	Argentina	1982 –	FR	France	1931 –
AT	Austra	1984 –	GB	United Kingdom	1970 –
CH	Switzerland	1970 –	HN	Honduras	1983 –
CL	Chile	1982 –	IT	Italy	1967 –
CO	Columbia	1977 –	MX	Mexico	1982 –
CR	Costa Rica	1983	NO	Norway	1976 –
CU	Cuba	1970	PA	Panama	1982 –
DD	German Dem. Rep.	1966 –	PE	Peru	1983 –
DE	Germany, Fed. Rep.	1877 –	PY	Paraguay	1983 –
		1945	SE	Sweden	1983 –
		1968 –	US	United States of	1920 –
EC	Ecuador	1982 –		America	
EP	European Patent	1985 –	UY	Uruguay	1979 –
	Office		VE	Venezuela	1982 –

- the following INPADOC services are available, Patent Classification Service (PCS), Patent Applicant Service (PAS), Patent Gazette (IPG), Patent Family Service (PFS), Patent Register Service (PRS).
- selective dissemination of information on Spanish patents by profiles.
- Spanish industrial property databases: –

CIBERPAT contains bibliographic data relating to more than 350,000 inventions published in Spain since 1968. It allows online access and is updated every 15 days.

INPAMAR includes 1,000,000 references, national trade marks, trade names, business signs and international trade marks registered in Spain. It can retrieve identities and similarities of all distinctive signs and is updated every 15 days.

SITADEX contains 500,000 records relating to the legal status of all industrial property rights registered in Spain, since 1979 for marks and since 1964 for patents in force.

VII Provincial libraries

The *Boletin Oficial de la Propiedad Industrial* is held at the following offices and libraries: –

- Delegaciones provinciales del Ministerio de Industria y Energia (provincial offices of the Ministry of Industry and Energy)
- Madrid, Biblioteca Nacional (National Library, Madrid)
- Madrid Hemeroteca National (National Library of Newspapers and Periodicals, Madrid)

INSTITUTO DE INFORMACION Y DOCUMENTACION EN   **ES**
CIENCIA Y TECNOLOGIA (ICYT)
c/Joaquín Costa 22
28002 MADRID
Tel. 34 1 2614808
Telex: 22628 CIDMDE
Telecopier 34 1 4113710

This Institute was set up in 1953 as part of the Higher Council for Scientific
Research under the Ministry of Education and Science. The Institute's main task
is information and documentation in the field of science and technology.

The Institute offers the following services

1. Retrospective bibliographical searches (BR), providing abstracts and/or
   bibliographical references on works published throughout the world in the
   field of science and technology, including patents.

   These are mainly on-line searches.

2. Selective dissemination of information (SDI) by means of computerized data
   banks containing international bibliographical data.

3. Copying facilities: the Institute, either through its own library or other
   libraries in Spain or abroad, is able to provide photocopies or microfilm
   copies of works published throughout the world.

4. Publications:

   - *Indice Español de Ciencia y Tecnología* (Index to Spanish Science and
     Technology), covering all papers published in Spanish scientific and tech-
     nical journals.
   - *Revista Española de Documentación* (Spanish Journal of Scientific Docu-
     mentation). Includes research papers, short notes and news on Informa-
     tion and Documentation Science.

FINLAND

FI

PATENT OFFICE

PATENTTI-JA REKISTERIHALLITUS
PATENT-OCH REGISTERSTYRELSEN
Albertinkatu 25
SF-00180 Helsinki 18
Tel.: (90) 69531

I   General information

The Finnish Patent Office is a body under Ministry of Commerce and Industry.
It is an examining patent office and responsible for the grant of patents and for
the registration of trade marks and designs in Finland.

II   Patent documents (Finnish-Swedish)

Patent applications (A)

Patent applications are open to public inspection 18 months after the day of filing or priority date, but they are not printed. Copies of such applications are
supplied by the Patent Office on request, to foreign subscribers through the official agent.

Kuulutusjulkaisu − Utläggningsskrift (B)

The specification of an examined and accepted patent application published to
allow opposition by third parties.

   Issued since 1968 (No 40 001 −)

Patenttijulkaisu − Patentskrift (C)

The specification of a granted patent. The specification is only reprinted when
the patent is granted, if it has been amended, e.g. due to an opposition.

   1900 (No 932) − 1942 (No 19 871): abstracts
   1943 (No 19 872): printed specifications.

[B] (11) **KUULUTUSJULKAISU**
UTLÄGGNINGSSKRIFT 72621

(45)

(51) Kv.lk.⁴/Int.Cl.⁴ H 01 R 4/30

SUOMI—FINLAND

(FI)

Patentti- ja rekisterihallitus
Patent- och registerstyrelsen

(21) Patenttihakemus — Patentansökning 860952
(22) Hakemispäivä — Ansökningsdag 07.03.86
(23) Alkupäivä — Giltighetsdag 07.03.86
(41) Tullut julkiseksi — Blivit offentlig

(44) Nähtäväksipanon ja kuul.julkaisun pvm. —
Ansökan utlagd och utl.skriften publicerad 27.02.87

(86) Kv. hakemus — Int. ansökan

(32)(33)(31) Pyydetty etuoikeus — Begärd prioritet

(71)(72) Reino Saarinen, Nummen Puistokatu 7, 20540 Turku, Suomi-Finland(FI)
(74) Oy Kolster Ab

(54) Maadoituspuristin hitsauslaitetta varten -
Jordningsklämma för svetsaggregat

(57) Tiivistelmä

Keksinnön kohteena on maadoituspuristin hitsauslaitetta varten. Maadoituspuristin käsittää rungossa (1) kiertyvän kiinnitysruuvin (2) ja runkoon (1) muodostetun vasta-
leuan (5), jotka yhdessä muodostavat ruuvi-
puristimen, jolla maadoituspuristin kiinni-
tetään työkappaleeseen (9). Lisäksi maadoi-
tuspuristimessa on liitäntä maadoituskaapelin (7) kiinnittämiseksi runkoon (1). Tunne-
tuissa maadoituspuristimissa on rakenteelli-
sia puutteita, joista johtuen ne eivät ole
varmatoimisia eivätkä pysty kokoonsa nähden
siirtämään riittävästi virtaa. Puutteet joh-
tuvat siitä, ettei puristimissa ole riittävästi huomiota sähkövirran varman kulun ta-
kaavaa kunnollista kontaktia työkappaleen
ja maadoituskaapelin välillä. Näiden ongel-
mien poistamiseksi käsittää maadoituskaapelin (7) liitäntä vastaleukaan (5) kohtisuo-
rasti kiinnitysruuvin (2) liikesuuntaa vas-
taan muodostetun ontelon (6), ja kiinnitys-
elimet (14, 15), jotka lukitsevat sekä maadoituskaapelin paljaan osan (12) että kaa-
pelieristeen (13) onteloon (6).

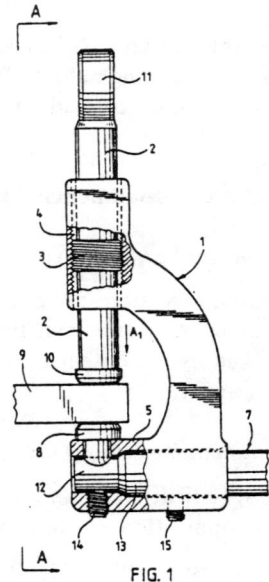

FIG. 1

In addition copies of the above mentioned documents are issued as 8-up aperture cards from 1972 – 83 and since 1984 on microfiches.

The front page of the documents and the punched information on the aperture cards present all bibliographic data registered by the Patent Office until the publication of the document, in accordance with WIPO standard ST 9.

III Official gazette and other publications (Finnish-Swedish)

Patenttilehti – Patenttidning   1889 –   monthly
(Finnish Official Patent Gazette)

Table of contents
○ *Julkiseksi tulleita patentti-*   ○ Patent applications open to public
   hakemuksia Patentansökningar   inspection
   som blivit offentliga
○ *Nähtäväksi pantuja patentti-*   ○ Examined, accepted and published
   hakemuksia – Utlagda patent-   ○ patent applications
○ *ansökningar*   ○ Granted patents
   Myönnettyjä patentteja –
   Meddelade patent
○ *Julkiseksi tulleita ja sen jälkeen*   ○ Publicly accessible applications
   peruutettuja, sillensä jätettyjä tai   withdrawn, abandoned or refused
   hylättyjä patenttihakemuksia –
   Patentansökningar som blivit
   offentliga och därefter återtagits,
   avskrivits eller avslagits
○ *Luovutettuja patenttihakemuksia –* ○ Assigned patent applications
   Överlåtna patentansökningar
○ *Luovutettuja patentteja –*   ○ Assigned patents
   Överlåtne patent
○ *Jälleen voimaansaatetut patentit –* ○ Restored patents
   Återupprättade patent
○ *Patentit, joiden voimassaoloaika*   ○ Expired patents
   on päättynyt – Patent vars
   giltighetstid utlöpt
○ *Patentit, jotka ovat lakanneet*   ○ Lapsed patents
   olemasta voimassa – Patent, vilka
   förfallit

Patenttilehti – Patenttidning No 13   1902 –   annual
(Annual register of Finnish patents)

Table of contents

○ Subject-matter index of issued patents.
○ Numerical list of issued patents indicating their classification.
○ List of patents which are valid.

○ *Patenttihakemusten viikkoluettelo – Veckoförteckning över patentan-sökningar*

Weekly list of filed patent applications.

IV Sources of supply and prices

Publications and photocopies are obtainable from the Finnish Patent Office, except the Finnish Official Patent Gazette, which is obtainable from:

VALTION PAINATUSKESKUS (the State Printing Centre)
PL 516 00101
Helsinki 10.

	Prices from 1. 1. 86 Fmk
Finnish Official Patent Gazette	
annual subscription	250. –
single copy	25. –
No 13 (annual register of patents)	25. –
	from 1. 1. 84
Weekly list of filed patent applications	
annual subscription	190. –
single copy	9. –
Finnish patent specifications	
single copy	9. –
Finnish printed examined applications (prices as for previous item)	
Photocopies of other documents, per page	0.50 – 3. –

The prices for patent documents and photocopies are exclusive of postage and commission.

V Register of legal status

The register is open to the public and contains full bibliographic details of applications, identification of agent, documentation received, fees paid and note of actions, decisions and deferments.

VI Public services

The reading room of the Finnish Patent Office is open to the public during office hours. On request specified groups of foreign patent documents may be
brought to the reading room for study. Some foreign patent specifications are
not available in classified sets and can only be obtained by number. Material
since 1960 has been reclassified by IPC but older patents are still classified by
the national system.

Books and periodicals on scientific, technical and legal subjects are available
through the reading room.

Patent documents in the library of the Finnish Patent Office

AT	Austria	specns	1963 –
		gazette	1963 –
AU	Australia	specns	1981 –
		gazette	1965 –
BE	Belgium	gazette	1968 –
BG	Bulgaria	specns	1968 –
		gazette	1968 –
CA	Canada	specns	1966 –
		gazette	1957 –
CH	Switzerland	specns	1944 –
		gazette	1962 –
CS	Czechoslovakia	specns	1961 –
		gazette	1964 –
CU	Cuba	gazette	1961 –
DD	German Democratic Republic	specns	1957 –
		gazette	1960 –
DE	Germany, Federal Republic of	specns	1877 –
		gazette	1877 –
DK	Denmark	specns	1900 –
		gazette	1902 –
EP	European Patent Office	specns	1978 –
FI	Finland	specns	1943 –
		gazette	1889 –
FR	France	specns	1951 –
		gazette	1951 –
GB	United Kingdom	specns	1953 –
		gazette	1931 –
HU	Hungary	specns	1985 –
		gazette	1973 –
IE	Ireland	gazette	1972 –
IT	Italy	specns	1977 –
JP	Japan	specns	1952 –
NL	Netherlands	specns	1964 –
		gazette	1964 –

NO	Norway	specns	1892 –
		gazette	1926 – 43; 1952 –
PL	Poland	specns	1934 –
		gazette	1953 –
SE	Sweden	specns	1885 –
		gazette	1907 – 28, 1949 –
SU	Soviet Union	specns	1952 –
		gazette	1958 –
US	United States of America	specns	1946 –
		gazette	1872 –
WO	WIPO-PCT	specns	1978 –
YU	Yugoslavia	gazette	1956 –
ZA	South Africa	gazette	1973 –

Special services

The Service Section of the Finnish Patent Office offers the services listed below. The person requesting an examination should specify a maximum price based on the number of hours of work involved.

— Novelty searches based on a short description of an invention. The report will refer to disclosures in patent specifications considered relevant so that the feasibility of a patent application can be estimated.
— Investigations useful for the development of products. The report will give information indicated in the literature on recent technology in the field in question.
— Name search. The Service Section can advise whether a particular company or person has patents in certain countries or in specified technical fields.
— Investigations to ascertain whether a particular invention is the subject of a patent or a patent application in Finland.
— Patent family investigations. On request the Service Section will search for patents or pending applications belonging to a certain patent family.
— Technical assistance. If it is required to make searches in the reading room of the Patent Office assistance may be provided.

Routine searches take one to two weeks, novelty searches two to four weeks.

PATENT OFFICE

INSTITUT NATIONAL DE LA PROPRIÉTÉ INDUSTRIELLE (INPI)
26 bis, rue de Léningrad
F-75008 Paris
Tel: 42 94 52 52
Telex: 29 03 68 INPI
Telecopier: no (1) 42 93 59 30

I   General information

The *Institut National de la Propriété Industrielle* is a public organization under the authority of the *Ministère chargé de l'Industrie* (Minister of Industry) with administrative status and budgetary autonomy.

INPI, established more than 30 years ago, is responsible for: −

− the administration of the industrial property acts and regulations
− the granting of industrial property rights (patents, trade marks, designs and models)
− maintaining a register of commerce and companies

INPI is responsible for patent filing procedures, grant and publication.

The reading rooms for the consultation of patent documents, French and foreign, are open to the public. They are administered by the *Division de la documentation, des publications et de l'action régionale*. A library, administered by the *Bureau de la documentation juridique et technique*, is also open to the public.

Staff (total):	710
Engineers-examiners:	100
Division de la documentation:	108
(60 in Paris, 21 in Compiègne,	
27 in the regional centres)	
Bureau de la documentation	
juridique et technique:	15

II Patent documents

As required by the law of 2 January 1968 amended by the law of 13 July 1978, INPI publishes the following specifications printed on different coloured paper for ease of identification. Each publication is identified by an alpha-numeric code according to WIPO standard ST 16.

Applications (on green paper)

- *demandes de brevet d'invention* (A1)
     patent applications
- *demandes de certificat d'addition à un brevet d'invention* (A2)
     applications for a certificate of addition to a patent
- *demandes de certificat d'utilité* (A3)
     applications for a utility certificate
- *demandes de certificat d'addition à un certificat d'utilité* (A4)
     applications for a certificate of addition to a utility certificate

Patents (on white paper), second publication having already been published as an application

- *brevets d'invention* (B1)
     patents
- *certificats d'addition à un brevet d'invention* (B2)
     certificates of addition to a patent
- *certificats d'utilité* (B3)
     utility certificates
- *certificats d'addition à un certificat d'utilité* (B4)
     certificates of addition to a utility certificate.

From no. 1 559 501 (March 1969), the first specification to be produced by off-set printing, the INID code (ST 9) appears on the front page.

Historical. French patents may be divided into three groups:

  i. 1791 – 1902   These files contain mainly descriptions in manuscript of inventions. INPI has about 350 000 files in the reserve store at Compiègne where they are protected against deterioration. The files may be consulted there under supervision or, on request, in Paris. They are of special interest to students of the history of science. The manuscripts are arranged as follows:

    17 July 1791 to 8 October 1844: in alphabetical order of name of applicant
    9 October 1844 to 31 December 1901: in numerical order

 ii. 1902 – 69.   Complete texts of patents granted are published as printed specifications. For patents filed after 30 June 1956 there is also an abstract. This collection comprises about 20 000 bound volumes (about 1 400 000 specifications).

⑲ RÉPUBLIQUE FRANÇAISE

INSTITUT NATIONAL
DE LA PROPRIÉTÉ INDUSTRIELLE

PARIS

⑪ N° de publication : **2 548 829**
(à n'utiliser que pour les
commandes de reproduction)

㉑ N° d'enregistrement national : **83 11260**

㉕ Int Cl⁴ : H 01 J 35/04.

⑫ # DEMANDE DE BREVET D'INVENTION   A1

㉒ Date de dépôt : 6 juillet 1983.

㉚ Priorité :

㊸ Date de la mise à disposition du public de la demande : BOPI « Brevets » n° 2 du 11 janvier 1985.

㉠ Références à d'autres documents nationaux apparentés :

㉛ Demandeur(s) : THOMSON-CSF, société anonyme. — FR.

㉟ Inventeur(s) : Emile Gabbay et André Plessis.

㉝ Titulaire(s) :

㉞ Mandataire(s) : Philippe Guilguet.

㊴ Tube à rayons X à anode tournante muni d'un dispositif d'écoulement des charges.

㊲ L'invention concerne un tube à rayons X à anode tournante muni d'un dispositif d'écoulement des charges, dans lequel le courant anodique du tube 1 est établi entre un émetteur 21 couplé en rotation à l'anode tournante 6, et un collecteur 22 en position fixe. Ce dernier est électriquement relié à la haute tension + HT et l'émetteur 21 est électriquement relié à l'anode tournante 6, et comporte au moins une pièce P1 ayant une extrémité effilée E1 constituant une zone émissive d'électrons. Ces électrons sont captés par le collecteur 22 de manière à établir le courant anodique du tube 1 sans contact matériel entre l'émetteur 21 et le collecteur 22.

L'invention est applicable aux domaines de la radiologie en général, et notamment au diagnostic médical.

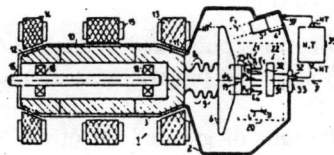

FR 2 548 829 - A1

95

iii. From 1969 (Law of 2 January 1968 amended). The first documents published according to the new Act were made available to the public from 29 August 1969 while publications under the old Law diminished.

These documents comprise:
– complete text of applications filed, on green paper
– complete text of granted patents, second publications, on white paper.

Numbering

i. Documents published prior to the Law of 2. 1. 68

Each patent in this series is allocated an issue number below 2 000 000. The first patent number published in 1902 is 317 502 and the last in this series to date is 1 605 584 and 96 692 for additions.

ii. Documents published under the new law

These have a national filing number comprised of the year of filing followed by the sequential number for that year, e.g. 85 1564, and a publication number in a continuous sequence starting at 2 000 001.

Brevets de médicaments (patents for medicines)

These patents, introduced by the Décret of 30 May 1960, which came into force on 1 June 1960, were published systematically from June 1960 with a separate numbering system: 1M, 2M, ...350M for patents and 1CAM, 2CAM, etc. for certificates of addition. The collection of patents for medicines comprises 8496 patents and 357 certificates of addition. Since the introduction of the new law these patents have been included in the normal patent system and are no longer published separately.

Micro-edition

Specifications published under the new law are also available on 35 mm, 8 up, film (about 250 patents on a 30 metre reel) and on aperture cards.

Translations of granted European patents effective in France and claims

In order to be effective in France translations of granted European patents must be provided within three months of the date of the announcement of grant in the *European Patent Bulletin*. INPI organises these translations which are made available to the public in Paris and at the regional centres in the form of aperture cards.

Also available are translations, on paper, of claims of European patent applications effective in France.

III Official gazette and other publications

Bulletin Officiel de la Propriété Industrielle (BOPI)   1884 –
Currently published in four editions: –

  BOPI – Brevets d'invention (Patents)   weekly
  BOPI – Marques de fabrique, de commerce ou de service (Trade marks and
        service marks)   weekly
  BOPI – Dessins et Modèles (Designs and models)   quarterly
  BOPI – Statistiques (Statistics)   annual

The following refers only to those editions concerning patents.

The edition entitled *Listes*, providing only bibliographic data on patents applied for and granted, was discontinued as a separate publication and from 23 April 1982 combined with *Abrégés du contenu technique de l'invention* to become *BOPI Brevets d'Invention (Abrégés et Listes)*. This publication has also been available on microfiches since 1983 by annual subscription.

BOPI – Brevets d'invention

Part I   *Première partie*

Abrégés du contenu technique de l'invention . . . .

Abstracts of published applications for patents, utility certificates and certificates of addition (1968 law)

The entries are arranged in numerical order of publication which corresponds also to the IPC order. Each entry comprises bibliographic data followed by an abstract of about ten lines usually accompanied by a drawing.

The publication of abstracts relating to patents granted under the 1844 law is now almost completed.

The *"Bulletin brevets"* is also available on microfiches on subscription. A full year comprises 250 microfiches which allows saving of space and improved searching.

Part II   *Deuxième partie*

Listes et tableaux relatifs aux demandes de brevets. . . . .

List of published applications for patents, utility certificates and certificates of addition.

1. Table alphabétique par noms des déposants et/ou titulaires	Name index to applicants and/or patentees
Table par matières	Subject index
Table par dates de priorité	Index arranged by priority date

It should be noted that from 23 April 1982 (*BOPI* No 16), the date on which the *Listes* and *Brevets d'Invention* were combined, the application numbers listed correspond to the abstracts published in the same issue. Consequently the indexes in Part II now relate to the abstracts in Part I.

2. *Liste des rapports de recherche rendus publics. Table numerique*

   Numerical list of search reports made available to the public

3. *Liste des seconds projets d'avis documentaires. Table numerique*

   Numerical list of second drafts of state of the art search reports

4. *Tableau des brevets pour lesquels un avis documentaire a été établi*

   List of patents for which a state of the art search report has been established

5. *Liste des brevets d'invention, certificats d'utilité, certificats d'addition délivrés*

   List of granted patents, utility certificates and certificates of addition

6. *Table de correspondance entre les numéros d'enregistrement nationaux et les numéros de publication*

   Concordance table – national filing number to publication number

7. *Tableau des inscriptions au Registre national des brevets*

   Table of entries in the national patent register

Part III – *Troisième partie*

Tableaux relatifs aux demandes de brevets et aux brevets européens produisant leurs effets en France
Lists of European patent applications and granted patents effective in France

1. *Tableau des brevets européens dont la traduction a été remise à INPI*

   List of European patents of which INPI has received the translation

2. *Tableau des brevets européens dont la traduction n'a pas été remise à INPI*

   List of European patents of which INPI has not received the translation

3. *Tableau des demandes de brevets européens pour lesquelles la traduction et, éventuellement, la traduction revisée des revendications ont été remises à INPI*

   List of European patent applications for which the translation and, eventually, the revised translation of the claims have been received by INPI

4. *Tableau des demandes de brevets européens transformées en demandes de brevets francais*

   List of European patent applications converted into applications for French patents

Other editions

BOPI Brevets spéciaux de medicaments 1961 – July 1973   weekly
BOPI Statistiques   1959 –   annual

Propriété industrielle – Bulletin documentaire (PIBD)
Oct. 1968 –   twice monthly

Produced by the *Bureau de la Documentation juridique et technique* and mainly concerned with industrial property right.

1. Textes officiels	Statutes and regulations
2. Doctrine – chroniques de l'étranger	Report on developments of industrial property law in foreign countries and on the international level
3. Jurisprudence . . .	Case law, French, EEC and EPO
4. Actualités et Informations	News and information, book reviews

Documents available for public inspection

The following may be consulted free of charge after announcement of publication in the official gazette and copies obtained.
– patent applications
– search reports
– granted patents

IV Sources of supply and prices

– Specifications of patent applications and granted patents, recent numbers

Direct from the Imprimerie National sales counter at INPI, Paris

By post (postage extra) from
Sevin
BP 637
41 rue Lille
59506 Douai Cedex
CCP 5707-41 R LILLE

– Annual subscriptions for patent specifications and official gazette also available from Sevin.
– BOPI – *Brevets d'Invention on microfiches*
from INPI, Division de la Documentation
– Reproduction of documents, photocopies, microfilms, paper copies from aperture cards, microfiches, microfilms
from INPI
A deposit account may be opened with INPI

- *PIBD*

Subscriptions and single copies from

La Documentation Francaise
124 rue Barbusse
93308 Aubervilliers Cedex
Tel: (1) 48 34 92 75

Single copies may also be purchased at INPI

- Aperture cards and microfilms of French patent specifications and European patent specifications in French from

Kodak
8 rue Villiot
75580 Paris Cedex 12

	Prices for 1986 FF
Specifications of French applications and granted patents, single copy** (details of subscriptions may be obtained from INPI)	13.00
Subscription for 8-up 35 mm microfilms of French patent applications	estimate can be supplied

BOPI – *Brevets d'Invention*

Normal edition (printed recto-verso)	
single copy**	45.00
annual subscription, France*	2100.00
annual subscription, abroad*	2500.00

Special edition (printed recto only)	
single copy**	70.00
annual subscription, France*	3150.00
annual subscription, abroad*	3750.00

Microfiche edition	
annual subscription, France*	1600.00
annual subscription, abroad*	1900.00

BOPI – *Statistiques*	50.00

PIBD

single copy**	22.50
annual subscription, France*	430.00
annual subscription, abroad*	550.00
annual subscription, abroad, by air	615.00

Reproduction on paper***
    all patent or legal documents   25.00
    translated abstract of European patent   14.00
    any other document of one or two pages****
    (trade mark, model, abstract, etc.)   6.00
    supplementary page   2.00

Reproduction on film***
    all patent documents   25.00
    any other document   estimate can be supplied
    copy of 16 mm film, 30 m reel,
    single price   250.00

   *   including postage
  **   excluding postage
 ***   the prices quoted include postage for France, increase of 20% for other countries
****   the reproductions are made from material available at INPI: paper, aperture cards, microfiches, microfilms

V   Legal documentation relating to patents

1. Register of legal status − *Registre National des Brevets (RNB)*

The *RNB* is a means of informing the public of the acts affecting patent rights. No entry can be made in the Register until the patent application has been announced in *BOPI-Brevets d'Invention* and against each entry is an indication of documents or action having effect against third parties.

The actual entry in the Register is published in BOPI.

A copy of the entries in the Register may be obtained against payment of a fee or, if there is no entry, a negative certificate is issued.

2. Industrial property card index

This index contains more than 95,000 references on the following aspects in the field of patents, trade marks, designs, unfair competition: −

− legislation in force in 150 states
− the doctrine of France, 60 states, European patent law
− French patent case law from 1837
− case law of the European Court of Justice
− decisions of the European Patent Office Appeal Court

3. JURINPI

Since the availability in 1974 of a legal database it is possible to obtain replies to questions of right or deed in 48 hours. Analysis may be provided of pertinent judicial decisions in France, the European Court of Justice and the European Patent Office. Photocopies of the decisions may be obtained from the legal library of INPI.

The cost is 200 F per enquiry or by subscription as follows: –

Up to 10 enquiries	1900 F
From 11 to 20 enquiries	3600 F
From 21 to 30 enquiries	5000 F

payable
– by cheques to l'Agent Comptable de l'INPI
– in cash at the INPI cash desk
– by deposit account

VI Public services

Legal and technical library

Open Monday – Friday 9.30 – 17.30

The library contains more than 20,000 works of which about 3000 are on the subject of French, foreign and international industrial property law and about 9000 on case law and economics.

The large collection of dictionaries and encyclopedia includes *Techniques de l'ingénieur, Kirk-Othmer Encyclopedia of Chemical Technology, Traité de chimie organique by Grignard, Encyclopédie des Sciences et des Techniques, Dictionnaire encyclopédique Quillet, Les normes francaises,* etc.

Public search room

Open Monday – Friday 9.30 – 17.30 and Saturday 9.00 – 12.00

There is no charge and no entry formalities. All the material is on open access: it must be consulted in the search room and may not be taken outside.

Card indexes

1. French patent abstracts arranged by class

   This index is made up from the abstracts published each week in the BOPI and includes abstracts of patents published in France since 1957 and all the abstracts, of applications filed since 1969. There are more than 1,500,000 abstracts and all IPC classes are included.

   After one year the paper entries are transferred to 16 mm film in 30 m cassettes. The period 1957 – 73 is cumulated.

This card index is also available at INPI regional centres, nine associated centres and several ARIST centres.

2. French patent abstracts arranged by name of applicant

This index starts at 1970 and is arranged by the most significant part of the applicant's name according to the AFNOR (*Association francaise de normalisation*) standard. It is also transferred annually on to 16 mm film and is available in regional centres.

3. Abstracts of European and PCT applications arranged by class

European patent and PCT specifications are not republished by INPI even if France is a designated country so INPI has prepared a card index of their abstracts arranged by class. The European patent abstracts are in the language in which the application was filed, i.e. English, German or French. The index is available on 16 mm microfilm from the date of entry into force of the EPC.

4. Abstracts in French of European patents

INPI has arranged for the translation into French of the abstracts of the 89% of European patents which are published in another language. A card index is available of abstracts in French of European patents arranged in numerical order. Each card relating to an abstract which was published in English or German carries the original abstract and the translation. This card index is also available on 16 mm microfilm.

Inpadoc

The following INPADOC services are available: –

> Patent Family Service (PFS)
> Patent Classification Service (PCS)
> Patent Applicant Service (PAS)
> Patent Inventor Service (PIS)
> Numerical Data Base (NDB)

Consultation of the PCS, also held at the regional centres of Lyon and Marseille, is free.

Copies on paper or microfilm of industrial property material held by INPI may be obtained by request in the reading rooms or by post and two self service photocopy machines are also available.

Patent specifications and official gazettes held at INPI either in Paris or in the reserve store at Compiègne

AR	Argentine	gazette	current year only
AT	Austria	specns	1899 –
		gazette	1957 –
AU	Australia	specns	1854 – 1967 (incomplete)
			1967 –
		gazette	1904 –
BE	Belgium	specns	1950 –
		gazette	1960 –
BG	Bulgaria	specns	1953 –
BR	Brazil	gazette	current year only
CA	Canada	specns	1949 –
		gazette	1904 –
CH	Switzerland	specns	1888 –
		gazette	1962 –
CN	China	specns	1985 –
		gazette	1985 –
CS	Czechoslovakia	specns	1915 –
		gazette	1921 – 39; 1952 –
CU	Cuba	gazette	current year only
DD	German Democratic Republic	specns	1974 –
		gazette	recent years only
DE	Germany, Federal Republic of	specns	1877 – 1945; 1949 –
		gazette	1877 – 1943; 1950 –
DK	Denmark	specns	1894 –
		gazette	1902 – 69; 1973 –
EP	European Patent Office	specns	1978 –
		gazette	1978 –
ES	Spain	specns	1981 –
		gazette	1886 – 1936; 1958 –
FI	Finland	specns	1944 –
		gazette	1952 –
FR	France	specns	1791 –
		gazette	1884 –
GB	United Kingdom	specns	1617 –
		gazette	1901 –
HU	Hungary	specns	1896 –
		gazette	1950 –
IE	Ireland	gazette	1960 –
IL	Israel	gazette	1982 –
IT	Italy	specns	1925 –
		gazette	1967 –
JP	Japan	specns	1950 –
		gazette	1960 –
		Pat. Abstr. of JP	
			1979 –
KR	Korea, Republic of	gazette	current year only
LU	Luxembourg	gazette	1967 –

MC	Monaco	gazette	1958 –
MX	Mexico	gazette	1983 –
NL	Netherlands	specns	1912 –
		gazette	1912 –
NO	Norway	specns	1893 –
		gazette	1911 –
NZ	New Zealand	specns	1982 –
		gazette	1909 –
OA	OAPI	specns	1966 –
		gazette	1967 –
PL	Poland	specns	1924 –
		gazette	1921 –
RO	Romania	specns	1957 –
		gazette	1969 –
SE	Sweden	specns	1899 –
		gazette	1885 –
SU	Soviet Union	specns	1896 – 1917; 1925 – 39; 1952 –
		gazette	1924 – 39; 1958 –
US	United States of America	specns	1871 –
		gazette	1901 –
VN	Vietnam, Socialist Republic of	specns	1984 –
		gazette	1986 –
WO	WIPO-PCT	specns	1978 –
		gazette	1978 –
YU	Yugoslavia	specns	1909 –
		gazette	recent years only

Special services

– Patent agent service

The *Compagnie des conseils en brevets d'invention* provides a free advisory service at INPI three times a week, on Tuesday, Wednesday and Thursday, 14.30 – 17.30.

– Independent applicants section

The majority of patent applications are made by large firms which, if not having their own industrial property service, are able to employ patent agents. Smaller firms and independent inventors themselves prepare and deposit their patent applications. To help such applicants the section *"Demandeurs Independants"* was formed at INPI.

It is composed of several engineers devoting nearly full time to the project each dealing with problems in a specialised subject field. The engineers are in direct contact with the applicants and inform them at three procedural stages – before deposit, at first technical examination and on reply to the search report.

The section does not advise on basic problems such as in the preparation of a legally defensible specification. It is intended only to enable applicants to comply with official requirements for the preparation of the text and does not in any way compete with the work of patent agents.

— Financial benefits for patent applicants

Benefits only to independent patent applicants (individual persons)
The cost of applying for a patent represents a considerable expense and arrangements have been made in France to relieve the financial pressure on needy inventors.

Any independent inventor who deposits a patent application may discharge the highest fee (for the state of the art report)
  — either paying by instalments over five years, without interest, if this is requested at the time of deposit.
  — or deferring payment: a maximum delay of eighteen months may be allowed if it is requested that the establishment of the state of the art report be also deferred by the same period of time.
It should be noted that all individual persons may deduct patent protection fees from their taxable income, particularly those fees relating to procedure.

Furthermore, independent persons domiciled in France whose income is below income tax level may be granted a general reduction on all fees imposed by INPI and also the free assistance of a patent agent during the INPI proceedings, provided that the invention is not obviously non-patentable.

Benefits available to all applicants

A reduction of 40% of the annual renewal fees is allowed on patents endorsed 'Licence of right'.

It should be noted that the owner of a potentially interesting patent may obtain certain grants from — *Aides à l'innovation de l'ANVAR (Agence Nationale de Valorisation de la Recherche)*, 43 rue Caumartin, 75436 Paris Cedex 09. Tel: (1) 42 66 93 10. Funds are provided by the *Fondation du brevet d'invention francais.*

If the invention is exploited, particularly if a licence is involved, the applicant may receive certain tax advantages.

— *Patent databases*

INPI-1 (FPAT) French patents

This database became available to the public in June 1980 and covers published French applications filed since 1969.

It contains more than 580,000 French patent documents and allows conversational access to their administrative and bibliographic data. Details of the state of the art report (*avis documentaire*) and the final search report (*rapport de recherche*) are included, i.e. patent numbers and references to non-patent literature. INPI-1 is updated weekly.

Additions in progress

- old law patents, applied for before 1969, and published since that date.
- patents for medicines (*brevets de médicaments*), about 9000 granted and published 1960 – 68. All the bibliographic data and the documents cited in the search reports have been included.
- old law patents, granted and published in 1968, 1967 and 1966 (corresponding to applications filed in 1966, 1965 and 1964 so that INPI-1 will cover the maximum protection period of 20 years).

INPI-2 (EPAT) European patents

Covers all published European patent applications and also Euro-PCT applications since their commencement in June 1978. It is the only database providing access to all the data on the European patent register. INPI-2 is updated weekly.

EDOC European documentation and patent families

Here are regrouped all the patent documentation of the European Patent Office covering the patents of seventeen countries and four organizations. The starting dates for the various authorities are: – DE 1877, GB-FR 1902, US-CH-BE-NL 1920, LU 1946, JP-AU-AT-CA-SE 1973, ES-GR-PT 1987 and OA-AP-EP-WO from their commencement.

Access to the data is by EPO classification (ECLA) and from 1968 by priority data. INPI-3, which provided information on patent families, is now integrated into EDOC. EDOC is updated monthly.

ECLA European Classification

This contains all of the 90,000 symbols of the European Patent Office classification which is derived from the IPC. Interrogation may be either by classification symbols or by keyword in English. ECLA is updated monthly.

Transinove

4000 licences from the principal industrial countries offered and requested.

JURINPI

French and European patent and trade mark case law. See para. V 3 for details.

Pharmaceutical chemistry database (in preparation)

Will contain data extracted from French, European and American patents in the field of pharmaceutical chemistry. Bibliographic descriptive and abstract data will be included as well as graphics (drawings and chemical formula presented in DARC format)

OAPI

Patents published by the *Organisation Africaine de Propriété Industrielle* (in preparation).

Rapid information service (*Service d'information rapide, SIR*)

This service was started concurrently with the introduction of the databases. It is located in the search room at INPI in Paris and operates also at the INPI regional centres. Subscribers may, by telephone or telex request, obtain in 12 hours any information contained in the INPI databases. SIR can also provide copies of relevant documents within 24 hours by post.

Some SIR subscribers have a terminal at their place of work but can call on INPI for the rapid supply of copies of documents. Requests for copies may also be made during interrogation of the databases INPI-1 and INPI-2 by using the appropriate code, the cost of the copies being deducted from the SIR account.

The majority of subscribers to SIR are industrial firms followed by patent agents and public services.

– Services to industry

In September 1985 the *Division de la promotion des services aux entreprises* was instituted. Its objectives are the production of two half yearly publications and the development of new services in conjunction with the databases.

Publications

Les techniques de demain: quelles seront-elles? d'ou viendront-elles?

Summarises the technical evolution in important fields such as energy, biotechnology and electronics.

Subscription (2 issues a year)	350 F
Single issue	200 F

Le brevet: clignotant de la technologie

Indicates the "hundred" technologies which are growing rapidly or are active with moderate growth.

Free on request

The data published is obtained through surveillance over a period of six months of French and European patents contained in the databases INPI-1 and INPI-2.

Technological surveillance

To determine
- technical evolution in certain fields
- principal lines of research at firms
- stages of progress of specified patent applications

Search reports on the state of the art

State of the art searches may be made at the request of any person.

- Standard searches of French patents applied
  for since January 1969 and European patents   1200 F
- Search restricted, for example, to a subject
  or applicant's name   600 F
- Searches which, because of difficulties in
  defining the problem, are not covered by the
  above tariffs, may be estimated separately.

INPI reserves the right to increase the value of the tariff or the estimate or to refuse a search, particularly if, as defined by the applicant, it appears likely to be unsuccessful.

The result of the search is presented in the form of a list of documents without comment or analysis.

Patent instruction

Association FORMEX

INPI, with the cooperation of the Association FORMEX, coordinates work in teaching establishments, research centres and industrial firms with help from specialists in industrial property matters.

FORMEX is an association for the promotion of industrial property created by the initiative of public and private organizations and universities to develop an interest in industrial property in industry, commerce, craft, research and education.

It is a non-profit making association with 51 patent agents, 31 engineers and 25 INPI personnel participating voluntarily. Courses and seminars are arranged at commercial and engineering colleges, universities and schools.

Teaching aids

An *educational (didactic) patent* in the form of a game was presented to the press in October 1986.

Black and white transparencies are available covering various aspects of industrial property at a cost of 1200 F for a set of 52.

The following *films* produced by INPI are available.

La guerre des idées   8 minutes

Innovation is essential to industry.

The world is constantly changing, technology evolves rapidly and innovation becomes a question of survival. It is necessary to anticipate this evolution, to know how to exploit our creativity, to commercialise and export our innovations. For this we must protect our inventions. The film is a warning, an incitement to industrial vigilance, a justification of creativity and innovation.

Format 16 mm — VHS U-MATIC

Eureka and C   10 minutes

Cartoon intended to alert the public to industrial property and the work of INPI. Panorama of important inventions and information on patentability.

Format VHS

B Comme brevets   5 minutes 50 seconds

Colour film with an electronic pencil visualising a patent. Why use a patent? How to obtain a patent? This film provides basic information on patents and attempts to reply to the sort of questions everybody asks about industrial property.

Format VHS

VII Provincial libraries

INPI regional centres

These are located at: —

BORDEAUX, 3 place Gabriel, 33075   Tel: (56) 90 91 28
LYON, 43 rue Raulin, 69007   (7) 872 59 42
MARSEILLE, 32 cours Pierre Puget, 13006   (91) 33 41 00
NANCY, 125 rue des Trois Maisons, 54000   (83) 30 57 60
NICE, rue Fernand Léger, Sophia-Antipolis 06560 Valbonne   (93) 65 37 37
RENNES, Centre d'Affaires Patton, 11,
rue Franz Heller 35000   (99) 38 16 68
STRASBOURG, 2 rue Brûlée 67000   (88) 32 25 44

The *work* of the INPI regional centres is as follows: −

− to make freely available to the public their collections of French patents, the official gazette BOPI and microfilmed abstracts arranged by class and name of applicant.
− to provide assistance in the use of IPC and information on industrial property rights and procedures.
− to ensure liaison with INPI Paris for
      searches on the various databases
      trade mark and trade name searches
      searches on INPADOC
      provide information on payment of patent renewal fees and certificate of registration in the register
      supply copies of patent and trade mark documents
− to promote industrial property in the various regions of France.
− to receive European patent and PCT applications and, since 1981, French patent and trade mark applications.
− to make use, by telephone, of the INPI rapid information service (SIR).

Holdings of INPI regional centres

FR −   specifications, official gazette BOPI, indexes to patents and trade marks, abstracts on microfilm
EP/WO − specifications, official gazette, abstracts on microfilm

Marques internationales and *US Official Gazette* (in some centres)
Technical journals − according to local requirements

Also at the following centres: −
      Lyon − DD, GB, US specifications
      Marseille − DD, DE, GB, US specifications
      Strasbourg − DD, DE specifications
      Bordeaux − ES official gazette

Documentation centres formed in association with the Chambre de Commerce et d'Industrie, the Chambres Regionales de Commerce et d'Industrie and certain archival or university services

BESANCON   ARIST-CRCI Franche-Comte
      30, avenue Carnot
      25043 Besancon Cedex
      Tel: 81 80 41 11

CAEN   Bibliothèque de l'Université de Caen
      Section sciences
      14032 Caen Cedex
      Tel: 31 94 81 40

CLERMONT-FERRAND	Bibliothèque de l'Université Section sciences Campus Universitaire des Cezeaux 63170 Aubiere Tel: 73 26 42 46
DIJON	ARIST-CRCI de Bourgogne 68, rue Chevreul – BP 209 21006 Dijon Cedex Tel: 80 67 33 25 Bureaux: 40, rue Gambetta, 21000 Dijon
LILLE	ARIST-CRCI Nord-pas-de Calais 77, rue Nationale 59800 Tel: 20 57 60 77
MONTPELLIER	ARIST-CRCI Languedoc-Roussillon Les Tonnelles, Bât. E, Entrée 3 131, avenue de Lodève BP 6076 Saint-Clément 34030 Montpellier Cedex Tel: 67 75 75 49
NANTES	SIDETEC-INNOVATION-CRCI des pays de la Loire et ARIST 29, quai de Versailles – BP 1100 44000 Nantes R.P. Tel: 40 48 64 09
ROUEN	Archives du Departement de la Seine-Maritime Cours Clémenceau 76036 Rouen Cedex Tel: 35 62 81 88 Exts. 621 – 631
TOULOUSE	ARIST-CRCI Midi-Pyrenées Parc Industriel Aéroportuaire de Toulouse-Blagnac 5, rue Dieudonné Costes – BP 32 31700 Blagnac Tel: 61 71 11 71

The work of these associate centres is, in general, the same as that of the INPI centres. They also have a collection of French patents, BOPI, indexes and microfilmed abridgments.

PATENT OFFICE

UNITED KINGDOM PATENT OFFICE
State House
66 – 71 High Holborn
London WC1R 4TP

Tel. 01-831 2525
Telex (Sale Branch) 896 348 PATENT OFF ORPTN

I   General information

The United Kingdom Patent Office is an administrative body placed under the Department of Trade. It is an examining patent office and responsible for the grant of patents in the United Kingdom and Northern Ireland.

Proposals exist to separate the Patent Office and establish it as a non-departmental, non-Crown public body. This would take the legal form of a body corporate, the members of which would constitute a Board. The Board members would be appointed by the Secretary of State and would comprise a Chairman as well as the Comptroller-General of Patents, Designs and Trade Marks together with other executive and non-executive directors. The staff of the Office would operate under the direction of the Comptroller. Neither Board members nor the staff would be civil servants.

Staff (total):   1210
Patent examiners: 325

II   Patent documents

A new Patents Act 1977 came into force on 1 June 1978 and introduced a deferred examination system. Applications upon which a complete specification had been filed prior to 1 June 1978 were examined under the previous legislation, the Patents Act 1949, and the complete specification was printed and published after acceptance.

For applications under the 1977 Act, an early publication (A) of the application together with a search report is made approximately 18 months after its priority

(12) **UK Patent Application** (19) **GB** (11) **2 165 007 A**

(43) Application published 3 Apr 1986

(21) Application No 8036344

(22) Date of filing 11 Nov 1980

(71) Applicant
Rolls-Royce Limited (United Kingdom),
65 Buckingham Gate London SW1E 6AT

(72) Inventor
Robert Rudolph Moritz

(74) Agent and/or address for service
J. C. Purcell, Company Patents & Licensing Dept., Rolls-
Royce Limited, P.O. Box 31 Moor Lane, Derby, DE2 8BJ

(51) INT CL⁴
F01D 11/00 F02C 3/06 7/18

(52) Domestic classification
F1T B2R
F1G 5C3A 6

(56) Documents cited
GB A   2032531

(58) Field of search
F1T
F1G

(54) **Rotor and stator assembly for a gas turbine engine**

(57) A rotor and stator assembly, particularly the turbine rotor (14, 17) and nozzle guide vane (30) assembly, of a gas turbine engine in which the inner platforms (29) of the vanes are provided with passageways (41, 43) into which is deflected radially outward flows (45, 46) from the face of the adjacent rotor discs (15, 18). Deflector means (24, 34) provide the necessary deflection. The passageways are shaped so as to convey the flow of air into the main gas flow of the engine with a minimum amount of disturbance, thus reducing the losses caused by these flows.

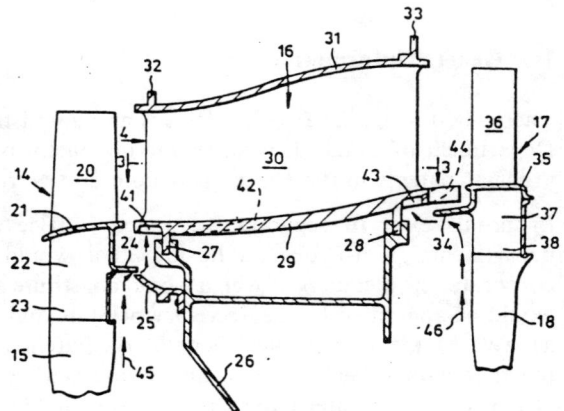

Fig. 2.

GB 2 165 007 A

The drawings originally filed were informal and the print here reproduced is taken from a later filed formal copy.

114

date. After substantive examination and grant the specification is published a second time (B) with the same serial number.

The chronological sequence of numbering British patents is: –

1617 to Sept. 1852:	a continuously numbered sequence to number 14,359.
Oct. 1852 to 1915:	annually numbered sequences in chronological order of application.
1916 to date:	a continuously numbered sequence from number 100,001 (1949 Act).
	a continuously numbered sequence from number 2 000 001.

Approximately 500 000 patent specifications were published before 1916 and for the annually numbered sequence both the number and the year of publication are needed for identification.

III Official gazette and other publications

Official Journal (Patents)   1854 –   weekly

Table of contents

Patent Office Publications
Official Notices
Recent judgements and decisions
Proceedings under the Patents Act 1977
– applications for patents
– applications withdrawn or refused
– applications published concurrently with the journal
– application number index of applications published
– name index of applications published
– patents granted
– subject matter index of patents granted
– patents and applications for patents in connection with which entries relating to assignments, transmissions, licences and the like have been made in the Register of Patents
– amendments printed
– patents granted under European Patent Convention
– patents revoked under European Patent Convention
– European patents ceased through non-payment of renewal fees
– patents ceased through non-payment of renewal fees
– patents expired

Proceedings under the Patents Act 1949
– significant stages of progress as above.

Classification Key

The United Kingdom has retained its own national patent classification system and this is indicated in the bibliographic data together with the IPC. It contains more than 400 "headings" (i.e. fields of subject matter) with detailed sub-classifying and indexing with more than 78000 coded terms within the headings. The headings are arranged into 40 divisions which, for publishing purposes, are grouped into 25 separate units, each unit containing one or more divisions according to the size of the division. The Key has been updated annually since 1 January 1979 when edition A was published.

A *"Universal Indexing Schedule" (UIS)* was introduced in 1983 providing indexing terms to record the uses and applications of inventions classified in other headings of the Key.

Abridgments/Abstracts

Abstracts of specifications published under the 1977 Act are sorted separately into the 25 units of the Classification Key and published weekly in pamphlet form.

Abridgment/Abstract volumes are published annually including both abridgments published under the 1949 Act and abstracts published under the 1977 Act.

Catchwords Index

This is for use in conjunction with the Classification Key and there is a separate volume for use with the *Universal Indexing Schedule (UIS)*

Name Index

A name index to applicants was published for each series of 25 000 specifications accepted under the 1949 Act. From 1 January 1979 an annual name index has been produced listing all applications published under a given name, both 1949 Act accepted specifications and 1977 Act A and B publications.

Reports of Patent, Design and Trade Mark Cases (RPCs)

Published irregularly since 1884; contains annual indexes of cases reported, defendants' names and cases referred to.

IV Source of supply and prices (from 1 January 1987)

	£
Specifications, 1949 Act, 1977 Act A publications	2.15
Specifications, 1977 Act B publications	3.15
Official Journal (Patents) per copy	4.30
Abstracts, per group unit (1987), subscription	57.00
Patents name index (1985)	80.00
Reports of Patent etc Cases, subscription	104.00

Up to date prices are to be found in the last issue of each month of the *Official Journal (Patents)* and a full price list is obtainable from the Sale Branch.

Deposit account system
A deposit account may be opened by customers in the United Kingdom and overseas for the supply of publications, public services and photocopies, but cannot be used for the payment of fees due to branches of the Patent Office other than the Sale Branch. The minimum amount that may be placed on deposit is £ 50 and replenishment payments should also be of £ 50.

Free publications, available from the Sale Branch

Patents, a source of technical information
Basic facts about patents for inventions in the United Kingdom
Introducing patents, a guide for inventors
How to prepare a UK patent application
Structure of the Classification Key
Classification and information retrieval services bulletin

All publications are available from: −

Sale Branch
Patent Office
Orpington
Kent BR5 3RD

V   Registers of legal status

A register is kept in the UK Patent Office in which are entered particulars of all patents in force and early publications. The register may be inspected by any person on payment of the prescribed fee and a certified copy of any entry may be obtained. There are some 250 000 patents at present in force including those granted under the European Patent Convention and designating the UK and the register is prima facie evidence of any matters entered in it in accordance with any provision of the Patents Acts 1949, 1957 and 1977.

VI Public Services

− Library
  Library facilities are provided by the British Library Science Reference and Information Service which has a separate entry.
− Selected patent specification service
  By standing order subscribers may be supplied with all UK patent specifications relating to any subject(s) as soon as they are published. The subjects must be expressed by means of codes of the current UK *Classification Key*.

There is no charge for this service other than the cost of the specifications supplied. It is available to deposit account holders only.

- Subject matter tabulation service

This service is similar to the one above except that lists of specification numbers are supplied instead of the actual specifications. This service is also only available to deposit account holders.

- Subject matter file lists

Each file list is a tabulation of the serial numbers of all specifications which have been allotted a selected classifying or indexing code or a logical combination of codes. It thus provides a ready means of identifying all specifications relating to a particular subject. Each file list is prepared to meet a specific order and there are the following three series: —

Series A — for each code mark in the *Classification Key* operative from specification no 960 001 to 1 000 000. These lists include all specifications from 1911 to no 1 000 000 indexed under the relevant mark.

Series C — for each code mark in the current and some old *Classification Keys*. These lists include specifications from a given year up to those currently published.

Series D — in respect of all headings for any combination of codes marks in the current and some old *Classification Keys* up to a maximum of 100 in any one heading; up to a maximum of 6 search patterns may be included on one list.

Further explanation and details of prices can be seen in the last issue of each month of the *Official Journal (Patents)*.

- File lists of foreign patents

Series C and D file lists are also available for files of foreign patents, mainly those of USA, Belgium, France, Federal Republic of Germany, the Netherlands and Switzerland, but for certain subject headings only.

- Search and Advisory Service

In an effort to increase the value to the public of the patent database, the United Kingdom Patent Office will, from Summer 1986, be introducing in stages new patent search services. An important aspect of these services will be state of the art searches, using the classified search files, arranged primarily according to the United Kingdom Classification Key, and drawing on the expertise acquired by the examining staff in the course of their statutory duties. The state of the art services will also use online databases where appropriate. A range of services will be provided, including investigating what is known in a particular field of technology and searching for known solutions to a technical problem.

Additional services will be based on the UK Patents Online database (see below). Similarly, search and advisory services on trade marks are being offered.

For further information please contact the Office's Patent Classification Section.

118

- UK Patents Online

  Details of the UK and IPC classification allocated to UK patents as well as the basic bibliographic information will be available on-line from Pergamon InfoLine Ltd. All data items published on the front page of patent documents are covered other than abstracts. For further details contact the UK Patent Office or Pergamon InfoLine.
- Awareness Programme

  In furtherance of the government's policy objectives, the Patent Office has established a Publicity Unit whose purpose is to raise the general level of national awareness of the importance of intellectual property, especially with regard to its commercial and industrial exploitation, and its place and function in strategic corporate planning und development.

Among the means used to achieve these purposes are: −

a seminar/lecture service, tailored to the particular needs of the client.

a video film which lasts for 25 minutes; in VHS, Betamax and U-matic formats, which may be hired or purchased.

an extensive range of free literature which explains the full range of services offered by the Office.

For further information, please contact the Office's Publicity Unit.

THE BRITISH LIBRARY   GB
Science Reference and Information Service (SRIS)
25 Southampton Buildings
Chancery Lane
London WC2A 1AW

Tel.: General enquiries   01-323 7494
      British and European patents   01-323 7919
      Foreign patents   01-323 7902
Telex: 266 959
Telecopier: 01-323 7930

I   General information

The Science Reference and Information Service (SRIS), now part of the Science
Technology and Industry division of The British Library, is based on the former
Patent Office Library which was opened in 1855 to encourage invention by making a comprehensive collection of relevant literature readily available to a wide
public. That remains the policy of SRIS; and the overall subject coverage has
since expanded to include literature in all branches of the technical and life
sciences and of technology.

Use of the library and its services (except photocopy supply, computer database
searching, and instructional courses) is free.

Much of the material is on open access with the patents and other industrial property publications in the following locations —

Main Library (Southampton Buildings)
   Open Monday to Friday 9.30 – 21.00; Saturday 10.00 – 13.00
Chancery House (opposite)
   Open Monday to Friday 9.30 – 17.30
Both closed on public holidays.

Staff (total):   305
Professional: 80

II   Holdings

Technical
SRIS has all GB specifications and associated publications and aims to have as
complete a holding as possible of the specifications and associated publications
(abstracts, name and subject indexes, patent classifications, etc) issued by all
other patenting authorities. SRIS also takes an extensive range of non-official
abstracts of patents, both those devoted exclusively to patents and others which
inlcude patents in their coverage of technical literature. It has online access to

120

a similarly wide range of computer databases. All holdings are taken in printed format if possible; however an increasing proportion of official industrial property publications is now only obtainable in microform.

Legal
Copies of industrial property legislation for GB and many other countries are held; and an up-to-date selection of reference works, monographs and journals covering the law, practice and history of industrial property is maintained. Complete files of officially published reports are held for GB industrial property cases heard from 1883 to date; non-official reports are held for significant earlier GB cases. Also held are reports of US, DE and EP cases. Reports of current period industrial property cases are accessible from certain of the computer databases.

Patent holdings in the British Library, Science Reference and Information Service

Note: This list is mainly of current items. All available earlier material is also held. Where an official gazette has changed title the earliest date held and the latest title have normally been listed.

AP ARIPO – *African Regional Industrial Property Organization:*
    Patent specification. 1985 – (1) –
    Patent and industrial design journal. 1985 –

AT AUSTRIA: *Patentamt*
    Patentschrift. 1899 – (1) –
    Österreichisches Patentblatt. 1899 –

AU AUSTRALIA: *Patent, Trade Marks and Designs Office.*
    Application (lapsed, refused or withdrawn before acceptance, having been laid OPI). 1959 – 1975.
    Application (laid OPI) 1986 –
    Patent specification 1904 – (1) –
    Australian official journal of patents, trade marks and designs 1904 –

BE BELGIUM: *Ministère des Affaires Économiques.*
    Brevet d'invention. 1950 – (493079) –
    Recueil des brevets d'invention. 1854 –

    BELGIUM: *Bureau Gevers.*
    Revue Gevers des brevets. 1961 –

BH BAHRAIN: *Ministry of Finance and National Economy.*
    Government gazette. 1956 – (139) –

BN BRUNEI: *Government Printing Department.*
    Government gazette, Part IV. 1970; 1978 –

BO BOLIVIA: *Ministerio de Industria Comercio y Turismo.*
    Gaceta oficial de Bolivia. Ano 3 – 1964 –

BR BRAZIL: *Instituto National da Propriedad Industrial.*
Pedido de Privilégio.
Patent application 1972 – 1974; PI 1975 –
Utility model application 1972; 1974 – 1976; MU 1977 –
Revista da propriedade industrial. 1972 –
(Diaro oficial 1922 – 72)

BS BAHAMAS: *Industrial Property Department.*
Official gazette. Supplement part III. 1889 –

BU BULGARIA: *Institut za izobreteniya i ratsionalizatsii.*
Opisanie na isobreteniya. 1952 –
Izobreteniya turgovski marki promishleni obraztsi 1964 –

BZ BELIZE: *Government Printer.*
Government gazette. 1923 –

CA CANADA: *Patent Office.*
Patent specification. 1948 – (445931) –
Patent Office record. 1873 – (1) –

CH SWITZERLAND: *Bundesamt für geistiges Eigentum.*
Patentschrift. 1888 – (1) –
Schweizerisches Patent-, Muster- und Markenblatt. 1889 –

CN CHINA: *Zhonhua Renmin Gongheguo Zhuanli Ju.*
Unexamined patent application. 1985 – (100001) –
Examined patent application. 1985 – (100001) –
Unexamined utility model application. 1985 – (200001) –
Faming zhuanli gongbao (Patent gazette). 1985 –
Shiyong xinxing zhuanli gongbao (Utility model gazette). 1985 –

CO COLUMBIA: *Superintendencia de Industria y Comercio.*
Gaceta de la propiedad industrial. 1958 –

CS CZECHOSLOVAKIA: *Urad Pro Vynalezy a objevy.*
Popis vynálezu . . . (Patents and authors' certificates).
1919 – (461) –
Vestnik. 1919 –

CU CUBA: *Oficina Nacional de Invenciones, Información Tecnica y Marcas.*
Boletin oficial. 1906 –

CY CYPRUS: *Dept. of Official Receiver and Registrar.*
Episémes efemeridas tes Demokratias. Parartima 5. Meros 2.
1961 –

DD GERMAN DEMOCRATIC REPUBLIC: *Amt für Erfindungs- und*
Patentwesen.
Patentschrift. 1951 – (1) –
Bekanntmachungen 1960 –

DE GERMANY, FEDERAL REPUBLIC OF; *Deutsches Patentamt.*
Offenlegungsschrift. 1968 – (1400003) –
Auslegeschrift. 1957 – 1985 (1000001 – 3038914)
Patentschrift. 1877 – (1) –
Patentblatt. 1877 –
Blatt für Patent-, Muster- und Zeichenwesen. 1948 –

GERMANY: *Wila Verlag für Wirtschaftswerbung Wilhelm Lampl KG.*
Auszüge aus den Offenlegungsschriften. 1968 –
Auszüge aus den Auslegeschriften/Patentschriften. 1955 –
Auszüge aus den Gebrauchsmustern. 1964 –

DK DENMARK: *Direktoratet for Patent – og Varemaerk – evæsenet.*
Patent. 1895 – (1) –
Fremlæggelsesskrift. 1968 – (111001) –
Dansk patenttidende. 1894 –

EP EUROPEAN PATENT OFFICE
European patent application. 1979 – (1) –
European patent specification. 1980 – (1) –
European patent bulletin. 1978 –
Official journal of the European Patent Office. 1978 –

Decisions – unpublished decisions of cases heard before the
Boards of Appeal of the EPO. 1980 –
– published decisions of cases heard before the Boards of Appeal
of the EPO. 1979 –

ES *SPAIN: Registro de la Propiedad Industrial.*
Patente de invencion. 1978 – 79; 1981 –
Modelo de utilidad. 1980 –
Boletin oficial de la propiedad industrial. 1886 –
Informacion tecnologica de patentes. 1981 –

FI FINLAND: *Patentti-ja Rekisterihallitus.*
Kuulutusjulkaisu. 1968 – (40001) –
Patentti. 1944 – (19901) –
Patenttilehti. 1889 –

FR FRANCE: *Institut National de la Propriété Industrielle*
Brevet d'invention. 1791 – (1) –
Demande de brevet d'invention. 1969 – (2000001) –
Brevet d'invention. 1969 – (2000001) –
Bulletin officiel de la propriété industrielle. 1884 –
Propriété industrielle bulletin documentaire. 1968 –
Brevets d'invention admis au régime de la licence de droit. 1980 –

GB UNITED KINGDOM: *Patent Office.*
Patent specification. 1617 – (1) –
Patent application. 1979 – (2000001) –
Patent 1981 – (2000001) –
Abridgments of patent specifications. 1617 –
Abstracts of patent applications. 1979 –
Official journal (patents) 1854 –
Reports of patent, design and trade mark cases. 1884 –

Decisions – transcripts of intellectual property cases heard before the Patent Office Court, the High Courts, and various other courts.

GG GUERNSEY: *Her Majesty's Greffier*
   Certificate of registration of patent. 1923 –

GR GREECE: *Ministère du Commerce.*
   Efémeris tés kubernéseos. Deltion emporikes kai biomékanikés idioktesias.
   1927 –

HK HONG KONG: *Government Printer.*
   Government gazette. Supplement no. 6. 1881 –

HU HUNGARY: *Országos Találmányi Hivatal.*
   Szabadalmi leírás. 1896 – (5792) –
   Abstract of patent specification. 1982 – (180101) –
   Szabadalmi közlöny es védjegyértesitö. 1896 –

IE IRISH REPUBLIC: *Patent Office.*
   Patent specification. 1928 – (1) –
   Official journal of industrial and commercial property. 1928 –

IL ISRAEL: *Patent Office.*
   Patents and designs journal. 1951 –

IN INDIA: *Patent Office.*
   Specification. 1912 – (1) –
   Gazette of India. Part 3, section 2. 1889 –

IT ITALY: *Ufficio Centrale dei Brevetti per Invenzione, Modelli e Marchi.*
   Brevetto per invenzione industriale. 1925 – (158965) –
   Bollettino dei brevetti per invenzione, modelli e marchi. 1902 –

JM JAMAICA: *Government Printer*
   Jamaica gazette. 1970 –
   (List of patents 1736 – 1977)

JO JORDAN: *Registrar of Patents, Designs and Trade Marks.*
   Official gazette. Supplement. 1930 –

JP JAPAN: *Tokkyo-chô (Patent Office)*
   Kôkai tokkyo kôhô. (Patent Application Gazette) 1971 –
   Tokkyo kôhô (Patent Gazette) 1889 – 1947; 1950 –
   Kokai jitsuyô shinan kôhô. (Utility Model Application Gazette)
   1905 – 1947; 1950 –
   Tokkyo hatsumei mesaishô. (Patented inventions) 1887 – 1956
   Shinketsu kôhô. (Reports of Legal decisions in cases involving
   Patent Rights). 1950 –
   Tokkyo-chô kôhô. (Patent Office Gazette) 1950 –
   Patent abstracts of Japan. 1977 –

KP KOREA, DEMOCRATIC PEOPLE'S REPUBLIC OF *(North)*
   Patent gazette. 1984 –

KR KOREA, REPUBLIC OF *(South): Office of Patents Administration.*
    Official gazette of the unexamined patent. 1983 –
    Patent specifications. 1954 – 55; 1961 –
    Official gazette of the unexamined utility model. 1983 –
    Utility model specifications. 1962 –
    Korean patent abstracts. 1979 –

LK SRI LANKA: *Department of the Registrar of Companies.*
    Gazette of the Democratic Socialist Republic of Sri Lanka.
        Part I. Section III, 1972 –
    (Ceylon Government Gazette 1881 – 1972)

LU LUXEMBOURG: *Recueil Administratif et Economique.*
    Mémorial. Journal officiel du Grand-Duché de Luxembourg.
        1916 – 25; 1946 – 59; 1963 –

MC MONACO: *Service de la Propriété Industrielle.*
    Annexe au journal de Monaco: protection de la propriété industrielle, littéraire et artistique. 1966 –

MT MALTA: *Dept of Trade.*
    Malta Government Gazette. 1900 –

MU MAURITIUS: *Ministry of Commerce and Industry.*
    Lists of letters patent granted. 1952 –

MW MALAWI: *Patent and Trade Marks Office.*
    Patent journal and trade marks journal. 1966 –

MX MEXICO: *Direccion General de Tecnologia, Invenciones y Marcas. Gaceta de invenciones y marcas. 1929 –*

MY MALAYSIA: *Registry of Trade Marks and Patents.*
    Warta Kerajaan. 1963 –

NL NETHERLANDS: *Bureau voor Industriële Eigendom.*
    Applications. 1930 – 1965
    Terinzagelegging. 1964 –
    Openbaarmaking. 1964 –
    Octrooi. 1913 –
    Industriële eigendom. 1912 –

NO NORWAY: *Styret for det Industriele Rettsvern.*
    Utlegningsskrift. 1968 – (115000) –
    Patent. 1892 – (2841) –
    Norsk tidende for det industrielle rettsvern. Patenter. 1886 –

NZ NEW ZEALAND: *Patent Office.*
    Patent specification. 1980 –
    Patent Office journal. 1912 –

OA OAPI *– Organisation Africaine de la Propriété Intellectuelle.*
    Brevet d'invention. 1979 – (3960) –
    Bulletin officiel 1966 – 70; 1977 –

PK PAKISTAN: *Patent Office*
Patent specification. 1958 – (107402) –
Gazette of Pakistan. Part V. Notifications etc issued by the Patent Office.
1955 –

PL POLAND: *Urząd Patentowy PRL.*
Opis patentowy. 1924 – (1) –
Biuletyn urzędu patentowego. 1973; 1976 –
Wiadomósci urzędu patentowego. 1924 – 39; 1946 –

PT PORTUGAL: *Instituto Nacional da Propriedade Industrial.*
Boletin da propriedade industrial. 1885 –

RO ROMANIA: *Oficiul de Stat Pentru Invenţii şi Marci.*
Descrierea invenţiei. 1957 – (39481) –
Buletin pentru invenţii şi marci. 1968 –

SE SWEDEN: *Kungliga Patent- och Registreringsverket.*
Utläggningsskrift. 1968 – (300001) –
Patentskrift. 1885 – (1-227869; 300001) –
Svensk patenttidning. 1968 –
(Svensk tidskrift for industriell rattsskydd. 1892 – 1974)

SG SINGAPORE: *Registry of Trade Marks and Patents.*
Government Gazette. 1946 –

SU SOVIET UNION: *Komitet SSSR po Delam Izobretenii i Otkrytii.*
Opisanie izobreteniya . . . (Patents and authors' certificates).
1924 – (1) –
Otkrytiya izobreteniya. 1924 –

TR TURKEY: *Sinaî Mülkiyet Dairesi Başkanligi.*
Resmi Sinaî Mülkiyet Gazetesi. 1931 –

TT TRINIDAD AND TOBAGO: *Registrar General's Department.*
Patent specification. 1887 –
Trinidad and Tobago gazette. 1966 –
(Trinidad Royal gazette. 1896 – 1962)

TW TAIWAN: *National Bureau of Standards.*
Official gazette – patents. 1974 –
(FORMOSA: Standardization monthly – patents 1954 – 73)

TZ TANZANIA: *Registrar of Trade Marks and Patents.*
Gazeti. 1965 –
(TANGANIKA: Tanganika gazette 1923 – 64)

US *UNITED STATES: Patent and Trademark Office.*
Patents. (Utility patents). 1790 –
Reissues. 1838 – (1) –
Design patents. 1931 – (1) –
Defensive publications. 1969 – 1986
Statutory invention registrations. 1985 – (Hl) –
Official gazette – patents. 1872 –

VN   VIETNAM, SOCIALIST REPUBLIC OF: *Ùy Ban Khòa Học Vàkỹ Thuât Nhã Nuoć.*
   Bàn mô tà sáng chê. 1984– (1)–
   Thông báo, 1984–

WO   PATENT COOPERATION TREATY: *World Intellectual Property Organization*
   PCT Application. 1978–
   PCT Gazette. 1978–

YU   YUGOSLAVIA: *Zavezni Zavod Za Patente.*
   Patentni glasnik. 1951–

ZA   SOUTH AFRICA: *Patent and Trade Marks Office.*
   Patent journal. 1948–

ZM   ZAMBIA: *Patents, Trade Marks and Designs Office.*
   Patent journal and trade marks journal. 1968–

ZW   ZIMBABWE *(formerly SOUTHERN RHODESIA): Patent and Trade Marks Office.*
   Patent and trade marks journal. 1960–

INPADOC – International Patent Documentation Center

   NDB   – *Numerical Data Base. 1968–*
   PCS   – *Patent Classification Service. 1968–*
   PAS   – *Patent Applicant Service. 1968–*
   PIS   – *Patent Inventor Service. 1968–*
   PAP   – *Patent Applicant Service Sorted by Priorities. 1968–*
   IPG   – *INPADOC Patent Gazette. Current issues*

DERWENT PUBLICATIONS LTD

World Patents Index –
   WPI Gazette. Current issues
   WPI Patent Number Family Index. 1974–
   WPI Accession Number Index. 1974–
   WPI Basic Patentee Index. 1974–
   WPI International Patent Classification Index. 1974–
   WPI Basic Priority Index. 1974–

World Patents Abstracts Journal. 1975–

BE   *Belgian Patents Abstracts. 1961–*
DE   *German Patents Abstracts. 1963–*
DE   *German Patents Gazette. 1968–*
EP   *European Patents Report. 1978–*
FR   *French Patents Abstracts. 1963–*
JP   *Japanese Patents Report. 1962–*
JP   *Japanese Patents Gazette. 1974–*

NL	*Netherlands Patents Report.* 1965 –
SU	*Soviet Inventions Illustrated.* 1962 –
US	*United States Patents Abstracts.* 1971 –
WO	*PCT Patents Report.* 1978 –

III Records of legal status: GB Patent Applications

Two series of manual registers recording the legal status of British patent applications are maintained for the convenience of those using the SRIS; they are unofficial and are compiled by library staff from information published in the GB *Official Journal (Patents)*. The information in these registers is continuous from 1914; the current series are known as the Applications and Stages of Progress registers respectively.

There is an Applications Register for each calendar year. Against the application number is entered the subsequent number of its published 'A' specification or the date of announcement in the *Official Journal (Patents)* when the application was withdrawn.

The Stages of Progress register is in continuous order of GB publication numbers and against each is entered the dates of announcement in the *Official Journal (Patents)* when the application was granted, then lapsed or expired.

IV Location of material

All GB, EP and PCT technical and legal publications on patents are located in the main library. The publications of all other patenting authorities are on the lower ground floor of Chancery House; preference there is given to publications from high-interest countries for the current search period. Other, including older, material is housed in one of the reserve locations but is accessible and obtainable upon request.

Non-official abstracting series covering foreign patents, such as those from Derwent Publications Ltd., are in Chancery House.

Technical and scientific abstracting journals, many of them covering patents, are located in the main library as are also the reference works, monographs and journals noted above.

V Information service

General

SRIS has staff at information points to answer enquiries and to assist visitors to the library to carry out patents searches, accessing computer databases where appropriate. Requests to SRIS for information may also be made by letter, telephone or telecopier. SRIS does not answer enquiries concerning industrial property matters which require professional advice or should be answered by the GB or other patenting authority.

Computer database searching

SRIS has access to a wide range of computer databases in its subject fields. Searches are conducted for anyone for charges which currently start at about £ 20 (eg. for a short search on one file).

Searches of patents databases are carried out by staff in the Industrial Property Section.

A free publication is available describing the SRIS Computer Search Service; this also includes a list of the major databases covering patents and trade marks, science and technology which can be accessed by SRIS.

VI Photocopies

In general, SRIS can supply photocopies of any material it holds, subject to copyright limitations and at a tariff of charges which relate to the several methods available for copying and delivery.

Photocopies of GB patent publications are ordinarily supplied by the UK Patent Office, not by SRIS.

SRIS can supply photocopies from its holdings of non-GB patent, design and trade mark publications normally without restriction and it will accept orders for such material where bibliographical details are not known exactly.

Orders can be sent to SRIS by post and also by electronic mail, telecopier, telex and telephone. If copies are required quicker than by airmail, delivery can be effected on request by telecopier.

Payment can be made using normal banking facilities or by opening a British Library account.

Initial enquiries about the SRIS current tariff of charges for photocopies and about the methods available for payment and delivering copies chould be made to the Photocopy Service, Science Reference and Information Service (address above).

VII UK Patents information network

Public libraries in the provincial libraries listed below, together with the SRIS in London, from the UK Patents Information Network providing public access to patent publications, assistance with searching and photocopies.

Full details of their holdings of patent publications and services offered may be obtained from the individual libraries or from the Industrial Property Section of the SRIS.

ABERDEEN
Central Library
Rosemont Viaduct
Aberdeen AB9 1GU

Tel: 0224 634622
Telex: 26587 MAN REF G
Telecopier: 0224 641985

BELFAST
Central Library
Royal Avenue
Belfast BT1 1EA

Tel: 0232 243233
Telex: 747359 BFTPL G

BIRMINGHAM
Central Library
Patents Department
Chamberlain Square
Birmingham B3 3HQ

Tel: 021 235 4537/8
Telex: 337655 BIRLIB G
Telecopier: 021 233 4458

BRADFORD
Central Library
Prince's Way
Bradford
W Yorks, BD1 1NN

Tel: 0274 753656
Telex: 51480 LIBRAD G

BRISTOL
Central Library
Commerce & Industry Dept
College Green
Bristol BS1 5TL

Tel: 0272 276121
Telex: 44200 AVCLBR G

COVENTRY
Lanchester Polytechnic
Much Park Street
Coventry CV1 2HF

Tel: 0203 24166

EDINBURGH
Central Library
George IV Bridge
Edinburgh EH1 1EG

Tel: 031 225 5584
Telex: 727143 EDICTY G

GLASGOW
Science & Technology Dept
Mitchell Library
North Street
Glasgow G3 7DN

Tel: 041 221 7030
Telex: 778732 LIBGLW G
Telecopier: 041 248 5027

KINGSTON-UPON-HULL
Central Library
Technical Library
Albion Street
Kingston-upon-Hull HU1 3TF

Tel: 0482 224040
Telex: 52211 HULLIB G

LEEDS
Patents Information Unit
Leeds Public Libraries
32 York Road
Leeds LS9 8TD

Tel: 0532 488747
Telex: 556237 LEEDS G

LEICESTER
Information Centre
Bishop Street
Leicester LE1 6AA

Tel: 0533 556699
Telex: 34307 LECLIB G
Telecopier: 0533 555435

LIVERPOOL
City Library
William Brown Street
Liverpool L3 8EW

Tel: 051 207 2147
Telex: 629500 LADSIR G
Telecopier: 051 207 1342

MANCHESTER
Central Library
Patents Library
St Peter's Square
Manchester M2 5PD

Tel: 061 236 9422
Telex: 667149 INFMRN G

MIDDLESBROUGH
Central Library
Victoria Square
Middlesbrough
Cleveland TS1 2AY

Tel: 0642 248155
Telex: 58439 CLEVEL G
Telecopier: 0642 248314

NEWCASTLE UPON TYNE
Central Library
PO Box IDX
Princess Square
Newcastle upon Tyne NE99 1DX

Tel: 091 261 7339
Telex: 53373 LINCLE G
Telecopier: 091 261 1435

NOTTINGHAM
County Library
Angel Row
Nottingham NG1 6HP

Tel: 0602 412121
Telex: 37662 NOTLIB G
Telecopier: 0602 412179

PLYMOUTH
Central Library
Drake Circus
Plymouth PL4 8AL

Tel: 0752 264675
Telex: 45578 WESLIB G
Telecopier: 0752 666213

PONTYPRIDD
Library
Polytechnic of Wales
Llantwit Road
Pontypridd
Mid Glamorgan CF37 1DL

Tel: 0443 405133

PORTSMOUTH
Central Library
Guildhall Square
Portsmouth PO1 2DX

Tel: 0705 819311
Telex: 86382 CENLIB G
Telecopier: 0705 839855

SHEFFIELD
Central Library
Surrey Street
Sheffield S1 1XZ

Tel: 0742 734742
Telex: 54243 SHFLIB G
Telecopier: 0742 735009

Patent and Trade Mark Searchers Group

Chairman   Mr W Edmondson
           8 Elm Tree Green
           Great Missenden
           Bucks
           England

Secretary Ms G Y Pugh
           Flat 2
           220 Randolph Avenue
           London W9

Founded

1973 as the PATMSA, Patent and Trade Mark Searchers Association

Reformed

October 1978 as the Patent and Trade Mark Searchers Group, a special interest group of the Institute of Information Scientists.

Renamed

October 1983 as the Patent and Trade Mark Group of the IIS.

Membership

Open to any individual who is a fellow, member, affiliate or student member of the Institute of Information Scientists. Other interested persons are permitted to become associates of the Group.

Objectives

- To provide a collective voice for patent and trade mark searchers in dealing with official bodies, both national and international.
- To advance the professional interests of patent and trade mark searchers.
- To advance the science and art of patent and trade mark searching.
- To secure the best facilities for the practice of patent and trade mark searching.

Activities

- The holding of meetings of members for discussion of all subjects relevant to the objectives of the Group.
- The arrangement of visits by members, e.g. to various libraries, patent offices, etc.
- The preparation and circulation of news letters.
- The organization of lectures, seminars and courses of instruction.
- Providing one representative for the FID/PD
- Providing Chairman/Secretary and two other representatives for the Patent Documentation and Classification Committee, which was set up by the group to meet representatives of the British Patent Office to discuss matters of documentation and classification.
- Providing at least one member for the SAC of the EPO.
- Delegating a member of the management committee to the IIS Council.
- Providing two representatives on the British Patent Office committee for Advanced Data Processing.
- Organising Sub-Committees to undertake specific tasks, these include: −
  a) Patent Office and Library Liaison Sub-Committee to meet officials of the Patent Office and Science Reference & Information Service (SRIS) as necessary, to discuss searching problems with the Patent Office and contents, cataloguing and layout matters with the Library.
  b) Publicity and Meetings Sub-Committee − to ensure that activities of the Group are publicised and to arrange meetings, speakers etc.
  c) Education Sub-Committee − to consider basic needs of the group in the field of education and to provide ideas for the meetings Sub-Committee.

Fees

A nominal fee of £ 3 is charged to Institute members; a fee of £ 6 is charged to non-Institute members.

PATENT OFFICE

MINISTRY OF INDUSTRY, ENERGY AND TECHNOLOGY
General Secretariat of Research and Technology
Patent Section
Michalacopoulou, 80
115.28 Athens
Tel: 01 777 5126 - 777 0377 - 770 0940

I General information

The Greek Patent Office is a part of the Ministry of Industry, Energy and Technology, General Secretariat of Research and Technology and is responsible for the granting of patents in Greece. It is temporarily located in the Ministry of Trade building at Canigos Square.

Patent applications are examined for formalities only. Greece has ratified the European Patent Convention and became a member in October 1986.

Trade marks are administered by the Ministry of Trade.

Staff (total):   21

II Patent documents

Greek patent specifications are not published but are laid open to public inspection on the day after grant which is no more than four months from the date of filing.

Photocopies of all Greek patent specifications, except those kept secret, are available at 10 drachmas per page if adequate notice is given.

III Official gazette

The Commercial and Industrial Property Bulletin issued monthly provides the following information on Greek patents and trade marks: –

- patents granted in numerical order with dates of filing, priority and grant, title of the invention, names and addresses of the applicant, representative and, where applicable, the inventor.
- assignments, licences, renunciations, forfeitures, restitutions and any other amendments to the status of the patent.

IV Source of supply and prices

Issues of the official gazette may be obtained from: –

National Printing Office
34 Kapodistriou Street
10432 Athens
Tel: 01-5248320
Telex: 22 32 11 YPET GR

Annual subscription, from 1 January 1986, 2100 drachmas.

V Register of legal status

Registers are kept at the Greek Patent Office for the following: –

- applications filed, in numerical order
- patents granted, in numerical order
- names of patentees, in alphabetical order
- inventions classified by subject matter

These registers are available to the public with the exception of that relating to applications filed which remain secret until grant.

There are also the following computerized indexes: –

- foreign organizations, companies, universities, etc.
- national organizations, companies, universities, etc.
- individual Greek patentees and inventors
- individual foreign patentees and inventors
- national organizations
- foreign organizations
- universities
- patents granted
- patents granted to patentees in the eleven regions of Greece

All the above are available to the public Monday to Friday, 12.00 to 14.30, or during office hours by appointment.

VI Public services

- At present there is no patent library open to the public.
- Specifications of Greek patents granted within the previous five years are on open access and those in remote storage are available at one day's notice.
- There is limited and incomplete coverage of technical journals and foreign official gazettes.
- No foreign patent specifications are held at present.

VII Provincial libraries

Regional offices of HOMMEX (Hellenic Organization of Small and Medium sized Enterprises and Handicrafts) hold a limited amount of patent related documentation, official gazette, regional indexes of patentees, etc.

IRELAND

PATENT OFFICE

PATENT OFFICE
45 Merrion Square
Dublin 2
Tel: 614144
Telefax: 765776

I   General information

The Patent Office of Ireland is an administrative body under the aegis of *an Roinn Tionscail agus Tráchtála*, i.e. the Department of Industry and Commerce. It is an examining patent office and is responsible for the grant of patents in the Republic of Ireland.

Staff (total):   89
Patent examiners: 15
Library staff: 4

II   Patent documents

On acceptance the complete specification is published and opposition may be filed within three months of the date of publication as defined in Section 18(2) of the Patents Act 1964. Application for sealing should be made within four months of publication.

Each application which has not been accepted is laid open to public inspection 18 months from the date of filing of the complete specification or from the earliest priority date, and is available, on request, at the public library of the Patent Office.

PATENT SPECIFICATION No. *(11)*

51616

Date of Application and Filing Complete
Specification: *(22)* 19 June, 1981
(21) No. 1368/81

Application made in: *(33)* United Kingdom (GB)
(31) No. 8020314   *(32)* 20 June, 1980

Complete Specification Published:
(44) 21st January, 1987

(51) Int. Cl. 4 B29C 7/01, B32B 31/30.

© **Government of Ireland** 1987

COMPLETE SPECIFICATION

(54) IMPROVEMENTS IN OR RELATING TO A METHOD AND
APPARATUS FOR PRODUCING MULTI-LAYER EXPANDED
FILMS.

PATENT APPLICATION BY *(71)* NATIONAL RESEARCH DEVELOPMENT
CORPORATION, A BRITISH CORPORATION ESTABLISHED BY STATUE,
OF 66-74 VICTORIA STREET, LONDON, SW1, ENGLAND.

Price 90p

138

III Official gazette

Official Journal of Industrial and Commercial Property   1928 –   every two
weeks

Table of contents

- List of patent applications received
- Complete specifications accepted with bibliographic details, abstract and
  drawing
- Applications abandoned, ceased, expired, etc.
- Complete specifications accepted but on which no patents will be sealed
- List of complete specifications open to public inspection

IV Source of supply and prices

Patent specifications are obtainable from the Patent Office, price IR£ 0.90 (Any
pre-September 1982 specifications remaining in stock cost IR£ 0.12 1/2.)

An annual subscription to the official gazette costs IR£ 95 and a single issue
costs IR£ 4, from January 1986. The gazette is obtainable from the

Government Publications Sale Office
Sun Alliance House
Molesworth Street
Dublin 2

V Register of legal status

Section 63 of the Patents Act 1964 decrees that a register of patents shall con-
tinue to be kept in which, inter alia, particulars of patents in force, of assignments, etc., are recorded.

The register is open to public inspection from Monday to Friday, 9.45 – 16.15,
upon payment of a fee of IR£ 6 (effective from 1 April 1986) per hour. The
Office accepts telephone enquiries about matters relating to entries in the
register.

VI Public services

- The library is open for public use between the hours of 9.45 and 16.15 from
  Monday to Friday. Literary works in the field of industrial and intellectual
  property are made available to the public free of charge. There is no loan
  facility. A photocopying service is operated in the library at a cost of
  IR£ 0.25 per page copied.

- Irish, British and European patent specifications as well as United States abridgments are held in the library. PCT specifications are available on microfiche. The specifications are arranged numerically, Irish abstracts are arranged by IPC and British patent abstracts by UK class order.
- Also held are official gazettes from Canada, Israel, United Kingdom, United States, the European Patent Office and PCT and *Patent Abstracts of Japan*.
- The subject coverage of books held in the public library includes works on intellectual property law and reports of patent cases in Ireland and in the United Kingdom.

PATENT OFFICE

L'UFFICIO CENTRALE BREVETTI
Via Molise 19
I-00187 Roma
Tel: (06) 46 54 53

I   General information

The Italian Patent Office is part of the *Direzione Generale Produzione Industriale* under the *Ministero dell'Industria del Commercio e dell'Artigianato* and is responsible for the grant of patents in Italy and San Marino. The applications are examined as to form, unity of invention and patentability; there is no examination as to novelty.

Staff (total including examiners and library personnel):   77

II   Patent documents

The current legislation covering patents of invention, *brevetti per invenzione industriali (A)*, is D.P.R. 30 June 1972, no. 540, art. 11. The patent specifications with drawings are laid open to public inspection eighteen months after their filing or priority date or after a delay of ninety days after the filing date if the applicant so requests. (R.D. 29 June 1939, no. 1127, art. 4)

The specifications are no longer printed but photocopies may be obtained on request. If required they may be certified "true copies".

The specifications for utility models, *modelli di utilità (U)*, are not printed but are laid open to public inspection under the same conditions as patents. Photocopies are also available.

III Official gazette

Bolletino dei brevetti per Invenzioni, Modelli e Marchi   monthly

Issued in three parts: –

Parte I   *Invenzioni industriali* (patents) 1902 –
Parte II   *Modelli industriali* (utility models) 1902 –
Parte III *Marchi d'impresa* (trade marks) 1913 –

IV Source of supply and price

Copies of patent specifications and the official gazette may be obtained from:

Istituto Poligrafico dello Stato
Direzione Commerciale
Piazza Verdi 10
Roma

Photocopies may be obtained from the Patent Office at a price of Lit. 100 per page.

V Register of legal status

The Patent Office holds the official files and original copies of patent applications on which are indicated details of their assignments or modifications of rights. The public may consult this collection and, on payment of a fee, obtain photocopies.

The price of each page of photocopy is Lit. 100 and this amount should be paid to the postal account 35596006 in the name of the Ufficio Centrale Brevetti.

At the Patent Office the public may also consult the collection of judgments on patent cases by the Court of Appeal.

VI Public services

The library is open Monday to Friday, 10.00 – 13.00

Patent specifications may be consulted without charge in the reading room as soon as they are laid open to public inspection. They are arranged in numerical order. The specifications are classified to IPC sub-class level and there is a classified card index available as well as an alphabetical name index.

The specifications are kept in store but may be made available quickly on request. Photocopy charges are as indicated in Para. V.

Database available to the public

The Patent Office computer contains the following details of each patent: −

1. Essential details of the application
2. Essential details of the patent
3. Identification of applicant and agent
4. Title
5. Inventor
6. Priority details
7. Fees paid

Searches may be made through access to the data recorded under 1, 2 and 3 for all types of patent and under 4 only for trade marks.

PATENT OFFICE

SERVICE DE LA PROPRIÉTÉ INDUSTRIELLE
Ministère de l'Economie nationale et des Classes moyennes
19-21, boulevard Royal
BP no. 97
Luxembourg
Tel: 4794-1 Ext. 315 or direct 4794-315

I   General information

The *Service de la Propriété Industrielle* is a public service under the Ministry of
Economic Affairs. It receives applications for patents and certificates of addition and, after formal examination, issues certificates of grant. It also receives
applications for European patents and international (PCT) applications.

Staff (total):   6

II   Patent documents

The Luxembourg patent documents are not printed. From the date of grant of
the certificate of protection the public have access to the patent specifications
and certificates of addition at the offices of the *Service des dossiers et des brevets*
and may obtain copies or certified true copies.

The patent or certificate of addition is granted after a minimum period of two
months from the date of filing. Delivery may be postponed at the request of the
applicant for a period not exceeding eighteen months from the same date.

The specifications may be submitted in either French or German but no official
translation into the other language is made.

III Official gazette

Mémorial – Journal officiel du Grand-Duché de Luxembourg, Recueil administratif et economique   quarterly

Brief details of patents granted with bibliographic data are published periodically in Section B – *Relevé officiel des brevets d'invention.*
All the patents granted which were filed during a three month period are listed in ascending order of their application numbers, which are the same as the patent numbers. This section is followed by a numerical table showing the relevant IPC sub-classes and a classified list arranged by IPC.

IV Source of supply and prices

The *Relevé officiel des brevets d'invention* and copies of Luxembourg patent documents may be ordered from the *Service de la Propriété industrielle* at Luxembourg.

	Price Lfrs
Memorial, Relevé. . .	80
Photocopies of documents available to the public, per page	15

Payable on receipt of an invoice to the receiving officer at

Administration de l'enregistrement et des domaines
Bureau des successions et de la taxe d'abonnement
(brevets d'invention)
Plateau du St Esprit
Luxembourg

The postal cheque number of the receiving officer is CCP 24373-26 (Luxembourg)

V Register of legal status

The *registre officiel des brevets d'invention* contains the main bibliographic data for Luxembourg patents and certificates of invention listed in chronological order of filing of the corresponding applications. Included is a list in alphabetical order of names of applicants with their corresponding patent numbers and grouped by year of application.

There is also a list of international patent applications designating the Grand-Duché which have been published internationally in accordance with Article 21 of the PCT. In addition there is a list of European patents designating the Grand-Duché which have been published in accordance with article 98 of the EPC.

There is no charge for a personal search of the register at the office.

VI Public services

The library of the Luxembourg industrial property service contains Luxembourg, European and PCT patent documents. The service also maintains a collection of periodicals, bulletins, yearbooks, commentaries and legal works in the industrial property field.

It is intended shortly to establish a terminal to provide access to the European patent register and the EPO databases FAMI, INVE and ECLA.

THE NETHERLANDS

NL

PATENT OFFICE

OCTROOIRAAD
Patentlaan 2
2288 EE Rijswijk
Postbus 5820

Tel: (070) 40 30 40
Telex: 33265 octrd nl

I General information

The Netherlands Patent Office is an administrative body with judicial powers placed under the Ministry of Economic Affairs. It is an examining patent office granting patents which are valid in the Kingdom of the Netherlands which at present consists of the Netherlands and Netherlands Antilles.

The European Patent Office (EPO), on the same site, and the Netherlands Patent Office use the same search files which are owned and managed by the EPO. The collection of patent documents arranged in numerical order in the library of the Netherlands Patent Office is available to the examiners.

Staff (total):   280
Patent examiners: 80
Library staff: 35

II Patent documents

Inventors wishing to file a patent application may obtain information from the patent office concerning the required procedure and formalities. A leaflet is available and also a list of registered patent attorneys (free of charge).

Terinzagelegging (A)

The unexamined application published and laid open to public inspection 18 months after the filing or priority date.

These have been published in two series:
Filed before 1.1.64   Nos 165 493 (1964) − 303 059 (1965)
Filed after 1.1.64   Nos 64 00 001 −

Octrooiraad

Nederland

(12) B **Openbaarmaking** (11) **181467**

(19) NL

(54) **Werkwijze voor het vervaardigen van een isotrope permanente magneet uit een legering van mangaan, aluminium en koolstof.**

(51) Int. Cl.[4]: H01F 1/04, C22F 1/16, C22C 22/00.

(71) Aanvrager(s): Matsushita Electric Industrial Co., Ltd. te Kadoma, Japan.

(74) Gem.: Ir. G.F. van der Beek c.s.
NEDERLANDSCH OCTROOIBUREAU
Joh. de Wittlaan 15
2517 JR 's-Gravenhage.

(21) Aanvrage Nr. 7413667.

(22) Ingediend 17 oktober 1974.

(32) Voorrang ingeroepen vanaf 19 oktober 1973.

(33) Land(en) van voorrang: Japan (JP).

(31) Nummer(s) van de voorrangsaanvrage(n): Nr. 118278/73.

(43) Ter inzage gelegd 22 april 1975.

(44) Openbaar gemaakt 16 maart 1987.

Openbaarmaking (B)

Specification of the examined and accepted application laid open to public inspection and published to allow opposition by third parties.

These specifications have been laid open to public inspection since 1912, published under their filing number from 1923 and under separate publication numbers from 1964 beginning with 120 001.

Octrooi (C)

Specification of the granted patent, issued since 1913. The INID code identifying the bibliographic data on the front page has been applied since January 1975.

Microforms. *Terinzageleggingen* and *Octrooien* published between 1 January 1975 and 31 December 1986 are available on 8-up aperture cards and subsequently on 16 mm film.

Availability of documents

Documents relating to patent applications laid open to public inspection may be inspected free of charge and copies obtained.

The availability is as follows: −

Documents relating to examination − on the day the examined specification is laid OPI

Documents relating to the final grant procedure if the patent has not been granted automatically − on the day the patent is published

Other documents − on the day the unexamined application is laid OPI or, if not then available, as soon as possible after receipt

III Official gazette and other publications

De Industriële Eigendom   1912 −   twice monthly

Table of contents

Deel I Terinzagelegging (on green paper)	Part I
A Ter inzage geledge aanvragen . . .	Patent applications laid open to public inspection
− *in nummervolgorde*	− in numerical order
− *in volgorde van de IPC*	− in order of IPC
− *in alfabetische volgorde van namen van aanvragers*	− alphabetically by name of applicant
C Overdracht van rechten . . .	Assignment of rights . . .

D Naamswijzigingen van aanvragers . . .	Modification of names . . .
F Terinzagelegging van vertalingen . . .Europese octrooiaanvragen	Translations into Dutch laid OPI of claims of European patent applications
G Verbeteringen met . . .	Corrections

Deel II Openbaarmaking (on white paper)	Part II
K Openbaarmaking van octrooiaanvragen	Publication of examined patent applications

Lists in numerical, IPC and alphabetical order

P Verlening van octrooien	Grant of patents

Lists in numerical, IPC and alphabetical order

R Mededelingen over openbaar gemaakte octrooiaanvragen	Further information on published applications
V Vervallen of . . .octrooien	Patents lapsed or annulled

Deel III Voor Nederland geldende Europese octrooien (on yellow paper)	European patents valid in the Netherlands
PE Verleende Europese octrooien	Granted European patents

Lists in numerical, IPC and alphabetical order

RE Nadere medelingen over Europese octrooien	Further information on European patents
SE Overdracht . . .	Assignments
TE Naamswijzigingen van houders	Changes of names of proprietors
VE Vervallen . . .	European patents lapsed or annulled

From 1964 to 1981 *De Industriële Eigendom* was published in two parts, *Deel I* — twice monthly and *Deel II* — monthly.

From 1982 there is a comprehensive contents list and explanation of the notes in English in each issue.

Computer produced multi-volume lists of patents laid open to public inspection but not yet granted and also of patents in force are available.

Jaarregister   1982 —

Annual alphabetical list of names of applicants or proprietors and classified list with IPC, patent number and reference to the official gazette in which the entry appeared. The index also covers European patents which have been mentioned in the official gazette. There are notes for use in English.

From 1964 to 1981 the index was issued quarterly as *Kwartaalregister*.

Bijblad bij de Industriële Eigendom 1933 – monthly

Contains official notices, reports and decisions on industrial property cases and articles on legal problems.

Indeling der Techniek (I.d.T.)

Classification key valid for the EPO search file until 1973 (*Interne Classificatie*).

IV Source of supply and prices

Publications and photocopies are obtainable from: –

Octrooiraad
Patentlaan 2
2288 EE Rijswijk

Prices for delivery within the Netherlands (from 1982)

	Fl
– Paper copies of patent specifications, per document	5. –
– Documents published after 1. 1. 1975 on 8-up aperture cards	
Granted patents	2. –
Applications published as filed	2.50
Subscription to IPC classes or names of patentees	
granted patents	1.50
applications published as filed	2. –
– Documents published after 1. 1. 1975 on	
35 mm 8-up roll film (provided on a monthly	
basis containing granted patents and examined	
applications which have not resulted in the	
automatic grant of a patent)	
Subscription for one year	600. –
– Photocopies of foreign patent documents, per page	1. –

De Industriële Eigendom

Annual subscription (including Jaarregister)	350, –
Single copy	15. –

Bijblad bij de Industriële Eigendom

Annual subscription	80. –
Single copy	8. –
Index	8. –

Indeling der Techniek
   Version valid for classified library file of
   Netherlands patents issued before July 1973   200. –
   Catchword index to Indeling der Techniek   40. –

V   Register of legal status

The following registers are kept at the Netherlands Patent Office: –
a) Filed patent applications in numerical order
b) Granted patents in numerical order
c) Deeds of assignment, licences, etc.
d) Names of applicants and patentees in alphabetical order

These registers may be inspected free of charge. Those registers mentioned under a, c and d must be kept secret until the patent application has been laid open to public inspection, i.e. 18 months after the filing or priority date.

VI Public services

Numerical and classified files of patent documents are on open access in the library which is open Monday to Friday 9.00 – 17.00.

Non-patent literature is available on request up to 16.00 and there are photocopying facilities.

All patent documents published by the *Octrooiraad* and the EPO are arranged by class, unbound in stationery boxes. Documents published before 1 July 1973 are arranged according to the *Indeling der Techniek (I.d.T.)*, as revised by the I.d.T. bulletin No. 67 of January 1972. Those published after that date are arranged by IPC.

GB classified abridgments from 1855 are held as well as the official gazettes of a number of countries containing abstracts arranged by class. Also available is the US patent classification subclass listing on microfilm.

Classification keys with their catchword indexes are held and also concordances IPC-I.d.T. (both senses at subclass level), US-IPC and DPK-IPC.

Visitors may, on request, consult books and periodicals in the collection of non-patent literature managed by the EPO for which there are alphabetical and subject (IPC) indexes. Here there are about 350 periodicals, of which about 120 are legal and the rest predominantly in the field of chemistry, including *Chemical Abstracts*. There are also about 10 000 books, 4000 legal and 2700 chemical. The legal works are mostly concerned with intellectual property.

Patent specifications and official gazettes held in the library of the Netherlands Patent Office

AR	Argentina	gazette	1923 – 61 (incomplete)
AT	Austria	specns	1899 –
		gazette	1878 – 94; 1899 –
AU	Australia	specns	1904 –
		gazette	1924 –
BE	Belgium	specns	1926 –
		gazette	1887 –
BG	Bulgaria	specns	1955 –
		gazette	1961 –
BR	Brazil	specns	1979 –
		gazette	1924 – 26;1946 – 52;1972 –
CA	Canada	specns	1949 –
		gazette	1949 –
CH	Switzerland	specns	1888 – 1921;1955 – 76
		gazette	1890 –
CN	China	specns	1985 –
		gazette	1986 –
CO	Colombia	gazette	1958 –
CS	Czechoslovakia	specns	1919 –
		gazette	1923 –
CU	Cuba	gazette	1961 –
DD	German Democratic Republic	specns	1951 –
		gazette	1960 –
DE	Germany, Federal Republic of	specns	1877 –
		gazette	1877 – 1945;1950 –
DK	Denmark	specns	1895 –
		gazette	1894 –
EP	European Patent Office	specns	1978 –
		gazette	1978 –
ES	Spain	specns	1986 –
		gazette	1886 – 95;1923 – 36;1960 –
FI	Finland	specns	1944 –
		gazette	1944 –
FR	France	specns	1791 –
		gazette	1888 –
GB	United Kingdom	specns	1630 –
		gazette	1855 –
GR	Greece	gazette	1966 –
HU	Hungary	specns	1911 –
		gazette	1911 –
IE	Ireland	specns	1929 – 76; 1983 –
		gazette	1928 –
IL	Israel	gazette	1951 –
IN	India	specns	1963 – 77
		gazette	1965 – (incomplete)

IT	Italy	specns	1886 –
		gazette	1902 – (incomplete)
JP	Japan	specns	1948 –
		gazette	1950 –
		Pat. Abstr. of JP	1977 –
KR	Korea	specns	1978 – 80
		gazette	1966 –
LU	Luxembourg	specns	1945 – (incomplete)
		gazette	1926 –
MX	Mexico	gazette	1903 – 73 (incomplete)
NL	Netherlands	specns	1913 –
		gazette	1912 –
NO	Norway	specns	1899 –
		gazette	1901 –
NZ	New Zealand	specns	1980 –
		gazette	1914 –
OA	OAPI	gazette	1966 – 70
PH	Philippines	gazette	1958 –
PK	Pakistan	gazette	1951 – 68
PL	Poland	specns	1924 –
		gazette	1924 – (incomplete)
PT	Portugal	gazette	1895 – 1917;1961 –
RO	Romania	specns	1957 –
		gazette	1959 –
SE	Sweden	specns	1885 –
		gazette	1932 –
SU	Soviet Union	specns	1897 – 1914; 1945 –
		gazette	1957 –
TR	Turkey	gazette	1931 –
TT	Trinidad	gazette	1923 – 62
TW	Taiwan	gazette	1978 –
US	United States of America	specns	1920 –
		gazette	1861 –
VE	Venezuela	gazette	1955 –
WO	WIPO-PCT	specns	1978 –
		gazette	1978 –
YU	Yugoslavia	gazette	1951 –
ZA	South africa	gazette	1937 – 40; 1948 –
ZR	Zaire	gazette	1950 – 59

Databases available to the public

Terminals may be used free of charge in the library to obtain information on published Netherlands and European patents. The public may order printouts (not free of charge) from the EPO databases INVE (classified) and FAMI (patent family).

DSM OCTROOIAFEDELING   NL

(DSM Patent Department)
Koestraat
Geleen

Open Monday – Friday 8.30 – 16.00
Please contact in advance. Tel: 04494-66701

A complete collection of the publications of the Netherlands Patent Office is available to the public, including: –

– examined and accepted Netherlands applications laid open to public inspection,
   a) in numerical order
   b) classified according to the classification system used by the Netherlands Patent Office
– unexamined Netherlands applications, published as filed, in numerical order only, partly on 35 mm roll film

Group lists of unexamined applications arranged according to the Netherlands Patent Office classification are available.

WERKGEMEENSCHAP OCTROOI-INFORMATIE NEDERLAND (WON)

NL

(Netherlands Association for Patent Information)

President   W G Vijvers
c/o Duphar B. V.
P.O. Box 900
NL-1380 DA Weesp

Secretariat   G K F van der Woud
c/o PTT-Patents Department
P.O. Box 430
NL-2260 AK Leidschendam
Telex: 31236 dnl nl

Founded: 1977

Members

Comprising: the Patent Office (Octrooiraad), semigovernmental organizations, private companies, patent bureaux, search organizations, individual persons.

Aims and tasks

- Promoting the improvement of the accessibility in the widest sense of patent documentation and patent information.
- Being the spokesman of the Netherlands community of patent documentation users on international meetings.
- Creating facilities for communication between Dutch patent documentation professionals.
- Maintaining contacts with sister organizations.
- Educational activities.

Activities

- Organizing meetings of all members.
- Publishing a Newsletter (about four times a year).
- Conveying information of a more general nature (desires, ideas, complaints) of the Dutch CPI-subscribers group to DERWENT and vice-versa).
- Representing the profession with respect to the Government in general and the Patent Office in particular. In that capacity taking part in the talks between the Patent Office Library and the public and in discussions within a committee advising the Dutch Government.
- Briefing our representatives in international organizations (FID/PD, EG-Working Group for Patent Documentation).

- Monitoring developments in existing patent documentation systems (use and revision of the IPC, EURONET databases etc.)
- Advocating the installation of new tools for patent documentation or the improvement of existing ones (PCT-documentation, legal status information, access to classified collections of patent literature etc.)

Membership

Public and corporate bodies and natural persons.

Annual fee: Dfl.   75 for personal membership.
             Dfl. 200 for institutional membership.

PATENT OFFICE

STYRET FOR DET INDUSTRIELLE RETTSVERN
Middelthunsgate 15b
Oslo
PO Box 8160 Dep.
N-0033 Oslo 1
Tel: +47 2 46 19 00
Telex: 19151 nopat n

I   General information

The Norwegian Patent Office is an administrative body under the Ministry of Industry and is responsible for the granting of patents in Norway. There are two examining departments, one for patents and the other for trade marks and designs.

Staff (total):   162
Patent examiners: 73
Library staff: 32

II   Patent documents

Patent applications (A)

Patent applications are laid open to public inspection 18 months after the filing or priority date but are not printed at this stage: photocopies of the specifications may be supplied by the Library on request.

Utlegningsskrift(B)   1968 –   115 000 –

Printed specification of an examined and accepted patent application published to allow opposition by third parties.

Patentskrift (C)   1893 –   2841 –

The specification of the granted patents is only reprinted if the *utlegningsskrift* has been amended following opposition or other cause.

NORGE
(19) [NO]

STYRET FOR DET
INDUSTRIELLE RETTSVERN

[B] (12) **UTLEGNINGSSKRIFT** (11) **NR.** 155901

(51) Int. Cl.⁴ F 23 D 11/34, B 05 B 17/06

(83)

(21) Patentsøknad nr. 803404
(22) Inngivelsesdag   12.11.80
(24) Løpedag   12.11.80
(62) Avdelt/utskilt fra søknad nr.

(71)(73) Søker/Patenthaver   SONO-TEK CORPORATION,
                              313 Main Mall,
                              Poughkeepsie, NY 12601,
                              USA.

(74) Fullmektig   A/S Oslo Patentkontor
                   Dr.ing. K. O. Berg, Oslo.

(86) Internasjonal søknad nr.   –
(86) Internasjonal inngivelsesdag –
(85) Videreføringsdag   –
(41) Alment tilgjengelig fra 14.05.81
(44) Utlegningsdag   09.03.87

(72) Oppfinner HARVEY L. BERGER,
               Poughkeepsie, NY,
               CHARLES R. BRANDOW,
               Highland, NY,
               USA.

(30) Prioritet begjært 13.11.79, 03.12.79, USA,
     nr. 93115, 95971.

(54) Oppfinnelsens benevnelse   ULTRALYD VÆSKEFORSTØVER SOM HAR EN AKSIELT
                                FORLØPENDE VÆSKEMATINGSPASSASJE.

(57) Sammendrag   Ultralydforstøver hvor væsken som mates til forstøv-
ningsoverflaten (22A) forløper aksielt gjennom forstøveren. Denne anordning muliggjør forbedret og forenk-
let kobling av væsketilførselsrøret (11B) til forstøveren. I en utførelsesform er et drivelement (54, 55,
56) anbragt mellom fremre og bakre hornseksjoner (58,
50) og en forstøvningsseksjon (16A) er koblet til front-
hornseksjonen (58). En passasje (52) strekker seg
aksielt gjennom den bakre seksjonen, drivorganet (54,
55, 56), frontseksjonen (58) og forstøvningsseksjonen
(16A) til en forstøvningsoverflate (22A). Drivorganet
innbefatter piezoelektriske elementer (54, 56) med ring-
formet utformning. Et rørformet element eller væske-
tilførselsrør (11B) opptas i nevnte passasje. I en
foretrukket utførelsesform innbefatter det rørformete
elementet eller væsketilførselsrøret en avkoblingshylse-
seksjon (70) og en trinnformet del tilpasset til å dan-
ne inngrep med en trinnformet del i den bakre seksjonen
ved kobling av det rørformete elementet eller væsketilførselsrøret til forstøveren for å trekke nevnte
fremre og bakre seksjoner sammen.

(56) Anførte publikasjoner

BRD (DE) off.skrift nr. 2734818,
Britisk (GB) patent nr. 951537.

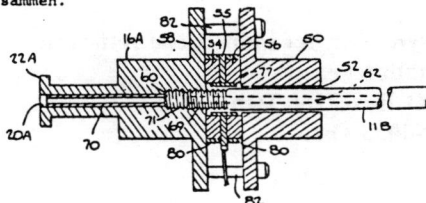

159

The INID code is used on the front page of the specification and microfiches are available as well as the printed edition.

III Official gazette and other publications

Norsk tidende for det industrielle rettsvern   Del I Patenter
1911 –   weekly

Table of contents

– *Alment tilgjengelige patentsøknader*	– Patent applications available to the public
– *Utlagte patentsøknader*	– Patent applications laid open to public inspection
– *Meddelte patenter*	– Granted patents
– *Tilbaketatte, avslåtte eller henlagte patentsøknader som er alment tilgjengelige*	– Withdrawn or refused applications available to the public
– *Patenter trådt ut av kraft*	– Lapsed or expired patents
– *Andre meddelelser*	– Miscellaneous announcements

Ukelister

Typed lists of applications filed are issued weekly by the Patent Office and copies may be obtained.

Register over norske patenter   1896 –

Annual name and classified index.

IV Source of supply and prices

Publications and photocopies are obtainable from:

Styret for det industrielle rettsvern
Biblioteket
PO Box 8160 Dep.
N-0033 Oslo 1

	Prices NKr
Norwegian official gazette	
annual subscription including index	200
single copy	10
Patent index	25
Printed patent applications	
annual subscription	800
single copy	10
Printed patent specifications	
annual subscription	800
single copy	10
Weekly list of applications filed	
annual subscription	500
Photocopies of other documents, per page	2

Postage is not included in these prices.

V Register of legal status

The register is available to the public and contains full bibliographic details of applications filed, identification of examiner, fees paid and notifications advised.

VI Public services

The reading room of the Norwegian Patent Office is open to the public on weekdays from 8.00 to 15.00 during the summer and from 8.00 to 15.45 during the winter.

Norwegian patent publications arranged in both numerical and class order are on open access in the reading room. There are also card indexes of pending Norwegian applications arranged by class and alphabetically by name of applicant and also of Norwegian patents granted from 1947 arranged by name of owner.

On request patent applications laid open to public inspection can be made available in the reading room.

The classified files are arranged as follows: –
  Nordic and German – partly German system (DPK), partly IPC
  French and British – mainly by Dutch system
  From 1975 all except US – IPC
  United States – US class, IPC from November 1981

In addition to the various classification manuals there are also available journals, books and dictionaries in scientific, technical and industrial property fields.

Patent specifications and official gazettes held in the library of the Norwegian Patent Office

AR	Argentina	gazette	1948 –
AT	Austria	specns	1899 –
		gazette	1899 –
AU	Australia	specns	1926 –
		gazette	1909 –
BE	Belgium	specns	1950 –
		gazette	1933 –
BG	Bulgaria	gazette	last 5 years
BR	Brazil	gazette	1972 –
CA	Canada	specns	1977 –
		gazette	1923 –
CH	Switzerland	specns	1907 –
		gazette	1908 –
CS	Czechoslovakia	gazette	1921 –
CU	Cuba	gazette	last 5 years
DD	German Democratic Republic	specns	1951 –
		gazette	1965 –
DE	Germany, Federal Republic of	specns	1877 –
		gazette	1877 –
DK	Denmark	specns	1894 –
		gazette	1913 –
EP	European Patent Office	specns	1978 –
		gazette	1978 –
ES	Spain	gazette	1965 – 76, 1983 –
FI	Finland	specns	1944 –
		gazette	1911 –
FR	France	specns	1902 –
		gazette	1903 –
GB	United Kingdom	specns	1884 –
		gazette	1893 –
HU	Hungary	gazette	1924 –
IE	Ireland	specns	1960 – 84
		gazette	1960 – 85
IL	Israel	gazette	1963 –
IS	Iceland	gazette	1978 –
IT	Italy	specns	1968 –
		gazette	last 3 years
JP	Japan	gazette	1977 –
MX	Mexico	gazette	last 5 years
NL	Netherlands	specns	1912 –
		gazette	1912 –

NO	Norway	specns	1893 –
		gazette	1886 –
NZ	New Zealand	gazette	1914 –
PL	Poland	gazette	1924 –
PT	Portugal	gazette	1962 –
SE	Sweden	specns	1885 –
		gazette	1896 –
SU	Soviet Union	specns	1955 –
		gazette	1929 –
TR	Turkey	gazette	last 5 years
TW	Taiwan	gazette	last 2 years
US	United States of America	specns	1904 –
		gazette	1876 –
WO	WIPO-PCT	specns	1979 –
		gazette	1978 –
YU	Yugoslavia	gazette	1951 –
ZA	South Africa	gazette	1965 –

Special services

The library has online access to INPADOC, SDC, DIALOG, Pergamon InfoLine and CAS and also subscribes to INPADOC indexes on microfiches.

Patent name and subject searches can be made and a current awareness service is also available.

VII Provincial libraries

Norges tekniske universitetsbibliotek
(The Technical University Library of Norway)
N-7034 Trondheim-NTH

This library holds Norwegian and some foreign patent publications. Norwegians and foreigners residing in Norway have free access to the public reading room and free borrowing facilities. A loan and photocopy service is offered to foreign libraries in accordance with usual library practice.

Some Norwegian public libraries subscribe to the Norwegian official gazette. In addition patent applications laid open to public inspection and/or granted patents are supplied to the public library of Bergen, the university libraries and some other research libraries.

PORTUGAL

PATENT OFFICE

INSTITUTO NACIONAL DA PROPRIEDADE INDUSTRIAL
Campo das Cebolas
1100 Lisboa
Tel.: 87 61 51/2/3
    87 11 01

I   General information

The *Instituto Nacional da Propriedade Industrial (INPI)* is a public institution, with administrative autonomy, attached to the Ministry of Industry and Commerce and responsible to the Secretary of State for Commerce. *INPI* is responsible for examination and grant of patents, utility models, industrial designs and models, trademarks, names and signs. After the current reorganization *INPI* will include various new services, such as a library, an international documentation service, a technical information office, etc.

Staff (total):   119

II   Patent documents

Patent specifications are not printed, but are made available to the public after publication of Patent Application abstracts.

Copies are available of all patent specifications in the publication phase.

The abstracts are published in the official gazette.

III Official gazette

Boletim da Propriedade Industrial   1885 –   monthly

Sumário	Table of Contents
– *Tribunais*	Decisions of the Courts
– *Patentes de invenção,*	Patents of invention,
– *Modelos de utilidade,*	Utility models,
Modelos Industriais e	Industrial models and
Desenhos industriais	Industrial designs
– *pedidos*	– patent applications
– *concessões*	– granted patents
– *revalidações*	– restored patents
– *notificações*	– notifications
– *caducidades*	– lapsed patents
– *averbamentos*	– additions
– *rectificação*	– amendments
– *Marcas nacionais*	National and international
e internacionais	trademarks
– *Nomes*	Names
– *Insígnias*	Signs

IV Source of supply and prices

Copies of all patents in the publication phase can be ordered from the *Instituto Nacional da Propriedade Industrial:*

	Prices Esc.
Patent specifications:	
Photocopies, per page	10. –
+ postage	
Official Gazette:	
Subscription in Portugal	2 500. –
+ postage	1 500. –
Spain	3 000. –
+ postage	1 500. –
other countries	4 000. –
+ postage	3 000. –

V   Register of legal status

The following indexes are publicly available at the *Instituto Nacional da Pro-priedade Industrial:*

- the number of the application (which becomes the number of the patent, when this is issued)
- the name and adress of the applicant
- the name and adress of any agent if the applicant is represented by an agent
- the name of the inventor
- the title of the invention
- the filing date of the application
- the priority data if a priority is claimed
- classification (IPC)
- status

With the reorganization of the *INPI*, all these bibliographic data, together with the data relating to the management of patents, will be processed by computer and constitute a publicly accessible data bank.

PATENT OFFICE

KUNGLIGA PATENT-OCH REGISTRERINGSVERKET
Valhallavägen 136
PO Box 5055
S-102 42 Stockholm
Tel.: (08) 782 2500
Telex.: 179 78 PATOREG-S

I   General information

The Swedish Patent Office is a public authority supervised by the Ministry of Industry (*Industridepartementet*) and has three departments viz. Patents, Limited Companies and Trade Marks.

It is an examining patent office and also serves as a PCT authority for the Nordic countries and for developing countries. Its Information Centre undertakes commissioned patent searches independent of any previous filings.

Staff (total):   688
Patent examiners: 140
Library staff: 42

II   Patent documents

Patent applications (A)

These are laid open to public inspection 18 months after the filing or priority date but are not printed at this stage: photocopies of the specifications may be supplied on request.

Utläggingsskrift (B)   1968 −   300 000 −

Printed specification of an examined and accepted patent application published to allow opposition by third parties.

SVERIGE (12) UTLÄGGNINGSSKRIFT [B] (21) 8104861-3

(19) SE

(51) Internationell klass 4 A61L 9/00

(44) Ansökan utlagd och utlägg-
ningsskriften publicerad 86-05-20
(41) Ansökan allmänt tillgänglig 83-02-18
(22) Patentansökan inkom 81-08-17
(24) Löpdag 81-08-17

(11) Publicerings-
nummer 444 890

Ansökan inkommen som:

PATENTVERKET

(62) Stamansökans nummer
(86) Internationell ingivningsdag
(86) Ingivningsdag för ansökan
om europeiskt patent
(30) Prioritetsuppgifter

☑ svensk patentansökan
☐ fullföljd internationell patentansökan
med nummer
☐ omvandlad europeisk patentansökan
med nummer

- -

(71) Sökande   Alfa-Laval AB, Box 500 147 00 Tumba SE
(72) Uppfinnare B A. Palm , Genarp
(74) Ombud   Clivemo I
(54) Benämning Anordning för upprätthållande av steril atmosför
                i tank med oxidationskänsligt innehåll

(56)Anförda publikationer: SE 385 843(A61L 3/00)

(57)Sammandrag:

I syfte att upprätthålla steril atmosfär i en lagringstank utan att tanken
tillåts kommunicera fritt med omgivningen samtidigt som risken för uppkomst
av farliga undertryck i tanken elimineras, försätts tanken med övertryck
relativt omgivande atmosfär genom att upptill anslutas till en luftledning
(4) genom vilken luft kontinuerligt matas till ett utlopp (5) via ett i
ledningen uppströms tanken anordnat sterilluftfilter (3) och via en i
ledningen nedströms tanken anordnad strypning (6).

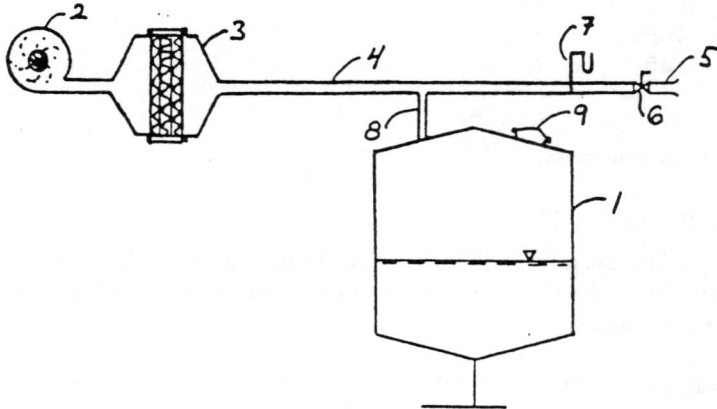

Siffrorna inom parentes anger internationell identifieringskod, INID-kod Bokstav inom klammer anger internationell dokumentkod

168

Patentskrift (C)   1885 – 1967   1-227 869

Before 1968 the examined and accepted patent application was laid open to public inspection but only the granted patent specification was printed.

Since 1968 the *utläggningsskrift* is only reprinted when the patent is granted if it has been amended following opposition or other cause.

The INID code has been used on the front page of specifications since 1968.

III Official gazette and other publications

Svensk Patenttidning   1885 –   weekly

Table of contents

○ *Inkomma patentansökningar*	○ Patent applications filed
○ *Allmänt tillgängliga patentansökningar*	○ Patent applications available to the public
○ *Utlagda patentansökningar*	○ Patent applications laid open to public inspection
○ *Meddelade patent*	○ Patents granted
○ *Återupprättade patent*	○ Restored patent
○ *Ansökningar om återupprättande av patent*	○ Applications for restoration of a patent
○ *Allmänt tillgängliga patentansökningar som slutbehandlats utan att leda till patent*	○ Patent applications available to the public which are finally decided without being granted
○ *Patent som upphört att gälla och avförts ur registret*	○ Expired patents deleted from the patent register
○ *Tillägg och rättelser*	○ Corrections

Bilaga till Svensk Patenttidning   (supplement to the official gazette)

Contains abstracts arranged by IPC of patent applications laid open to public inspection.

Svensk Patenttidnings kumulerade namnregister   monthly

Name index cumulated monthly and three monthly.

Årsförteckning över patent

Annual name and subject index.

IV Source of supply and prices

Publications and photocopies are obtainable from

KUNGLIGA PATENT- OCH REGISTRERINGSVERKET
P.O. Box 5055
S-102 42 Stockholm
Telex: 179 78 patoreg-s

	Prices (from July 1986) SKr
Swedish official gazette including the cumulated index of names	
annual subscription	700
Swedish official gazette	
annual subscription	400
single copy	10
Supplement to patent gazette, annual subscription rates	
A Human necessities	275
B Performing operations	700
C Chemistry and metallurgy	600
D Textiles and paper	200
E Fixed constructions	200
F Mechanical engineering, weapons, blasting	350
G Physics	325
H Electricity	325
All abstracts (*A-H*)	2500
Monthly index of names	
annual subscription	300

(from 1982)

Abstracts of patent applications	
per sheet	10
Index of Swedish patents (on microfiches)	
annual subscription	200
Published patent applications	
per copy	30
Patent specifications	
per copy	30
Photocopies of other documents	
per page	3

The prices for patent documents and photocopies are exclusive of postage.

V   Register of legal status (Diariet)

This record of filed applications is open to the public. It contains all the bibliographic data and also an index of documents filed, fees paid and decisions taken on the application. For the record of international applications, special rules are applied.

VI Public services

The main reading room of the Swedish Patent Office is open to the public from Monday to Friday from 10.00 to 16.30 and additionally on Tuesday from 17.00 to 19.00.

The following collections are accessible in this reading room: –

– classified set of printed Swedish patent documents.
– name indexes to accepted patent applications and granted patents.
– microfiches of Swedish applications laid open to public inspection.
– abstracts of unexamined Swedish applications laid open to public inspection, arranged by IPC (sub-class).
– official gazettes and abstracts from Austria, Belgium, Canada, France, United Kingdom, Netherlands, Scandinavian countries, Switzerland and the United States.
– abstracts of international publications arranged by IPC (sub-class).
– encyclopedias, dictionaries and other reference books.

On request patent documents, pending Swedish patent applications and other patent office publications, books and periodicals may be brought to the reading room for consultation.

In a second reading room classified collections of patent documents from Denmark, Norway and Finland as well as abstracts of European published applications are available to the public.

In another reading room outside the Patent Office building are patent specifications from the United States and the Federal Republic of Germany arranged by US classification and IPC respectively. This reading room is open from Monday to Wednesday from 9.00 to 16.00.

Patent documents in the library of the Swedish Patent Office

AR	Argentina	abstracts	1907 – (incomplete)
		gazette	1907 – (incomplete)
AT	Austria	specns	1899 –
		gazette	1899 –
AU	Australia	specns	1926 –
		gazette	1904 –

BE	Belgium	specns	1950 –
		gazette	1880 –
BG	Bulgaria	specns	1964 –
		gazette	1962 –
BR	Brazil	gazette	1972 –
CA	Canada	specns	1949 –
		gazette	1873 –
CH	Switzerland	specns	1888 –
		gazette	1889 –
CS	Czechoslovakia	specns	1919 –
		gazette	1921 –
CU	Cuba	abstracts	1961 –
		gazette	1961 –
DD	German Democratic Republic	specns	1958 –
		gazette	1960 –
DE	Germany, Federal Republic of	specns	1877 –
		gazette	1877 –
DK	Denmark	specns	1895 –
		gazette	1894 –
EP	European Patent Office	specns	1978 –
		gazette	1978 –
ES	Spain	gazette	1886 –
FI	Finland	specns	1944 –
		gazette	1899 –
FR	France	specns	1791 –
		gazette	1871 –
GB	United Kingdom	specns	1617 –
		gazette	1854 –
GR	Greece	gazette	1968 –
HU	Hungary	specns	1896 –
		gazette	1896 –
IE	Ireland	specns	1958 –
		gazette	1958 –
IL	Israel	abstracts	1963 –
		gazette	1968 –
IN	India	gazette	1953 –
IT	Italy	specns	1926 –
		gazette	1868 – (incomplete)
JP	Japan	specns	1952 –
MX	Mexico	abstracts	1938 –
		gazette	1903 –
NL	Netherlands	specns	1912 –
		gazette	1912 –
NO	Norway	specns	1886 –
		gazette	1886 –
PH	Philippines	abstracts	1963 –
		gazette	1963 –
PL	Poland	specns	1924 –
		gazette	1924 –

RO	Romania	specns	1957 –
		gazette	1921 – 32, 1966 –
SE	Sweden	specns	1885 –
		gazette	1885 –
SU	Soviet Union	specns	1863 – (incomplete)
		gazette	1924 – (incomplete)
TR	Turkey	abstracts	1931 –
		gazette	1931 –
US	United States of America	specns	1885 –
		plant specns	1940 – 43, 1955
		gazette	1850 –
WO	WIPO-PCT	specns	1978 –
		gazette	1978 –
YU	Yugoslavia	abstracts	1921 – (incomplete)
		gazette	1921 – (incomplete)
ZA	South Africa	abstracts	1953 –
		gazette	1955 –

Special services

The Information Centre of the Swedish Patent Office offers the following principal services to any person or company in any country. Fees are calculated according to the time spent and the enquirer may set an upper limit. The investigations are carried out by a member of the Patent Office technical staff and a report identifying retrieved documents and the scope of the search is sent to the enquirer.

1. Whether a solution to a particular problem is already known.
2. Novelty searches based on a short description of an invention. The report will identify relevant patent documents and will enable the enquirer to estimate whether
   a. a patent application should be filed, or
   b. an opposition should be filed in a foreign country, or
   c. the validity of a patent granted in a foreign country may be contested.
   The investigation is restricted to the technical content of the documents retrieved and the report does not comment or advise on the novelty or patentability of the invention.
3. Novelty searches for industrial firms proposing to launch a new product.
4. State of the art assessment from the patent literature in a specific subject field.
5. Indexes are available which will provide information on patent families, patents and pending applications in various countries in the name of an applicant or inventor and lists of patents in a certain IPC.
6. Alerting service to inform when applications are filed in a certain class and when a certain Swedish application is made available to the public or accepted and published.
7. Technical assistance by a specialist can be provided to establish a search profile and evaluate information from retrieved documents.

Investigations 1 – 4 can usually be completed within three weeks and replies involving consultation of the indexes (5) can be made directly by telephone/telex or promptly by letter.

VII Provincial libraries

Swedish patents are also available at: –

Malmö Stadsbibliotek
Regementsgatan
211 42 Malmö

Chalmers Tekniska Högskola
Biblioteket
Fack
420 20 Göteborg 5

WORLD INTELLECTUAL PROPERTY ORGANIZATION (WIPO)

34 chemin des Colombettes
CH-1211 Geneva
Switzerland
Tel: (022) 999 111
Telex: CH 223 76
Telecopier: (022) 335 428 (Groups 2 and 3)

I   General information

The international Bureau of WIPO, a specialized agency of the United Nations system since 1974, administers the Patent Cooperation Treaty (PCT) which was signed in Washington on 19 June 1970 and came into force on 24 January 1978. The following states have ratified or acceded to the PCT: –

Australia	Liechtenstein
Austria	Luxembourg
Barbados	Madagascar
Belgium	Malawi
Brazil	Mali
Bulgaria	Mauritania
Cameroon	Monaco
Central African Republic	Netherlands
Chad	Norway
Congo	Romania
Denmark	Senegal
Finland	Soviet Union
France	Sri Lanka
Gabon	Sudan
Germany, Federal Republic	Sweden
Hungary	Switzerland
Italy	Togo
Japan	United Kingdom
Korea, Dem. People's Rep.	United States
Korea, Rep. of	

When a PCT application is filed at a receiving office, normally the applicant's own national industrial property office, it is subjected to an international search by an International Searching Authority established under the PCT. A copy of the application and of the search report is sent to the International Bureau.

PCT

WORLD INTELLECTUAL PROPERTY ORGANIZATION
International Bureau

INTERNATIONAL APPLICATION PUBLISHED UNDER THE PATENT COOPERATION TREATY (PCT)

(51) International Patent Classification 4 :		(11) International Publication Number:	**WO 86/ 02059**
B65G 61/00, 47/90	A1	(43) International Publication Date:	10 April 1986 (10.04.86)

(21) International Application Number: PCT/EP85/00496

(22) International Filing Date: 24 September 1985 (24.09.85)

(31) Priority Application Number: 2/60510

(32) Priority Date: 28 September 1984 (28.09.84)

(33) Priority Country: BE

(71) Applicant *(for all designated States except US)*: BELL TELEPHONE MANUFACTURING COMPANY N.V. [BE/BE]; Francis Wellesplein 1, B-2018 Antwerp (BE).

(72) Inventor; and
(75) Inventor/Applicant *(for US only)* : VERHAEGHEN, Jacobus, Jan, Leon, Gerard [BE/BE]; Hendrik de Braeckeleerlaan, 37, B-2630 Aartselaar (BE).

(74) Agents: VERMEERSCH, Robert et al.; Bell Telephone Manufacturing Company N.V., Patent Department, Francis Wellesplein 1, B-2018 Antwerp (BE).

(81) Designated States: AT (European patent), BE (European patent), CH (European patent), DE (European patent), FR (European patent), GB (European patent), IT (European patent), JP, LU (European patent), NL (European patent), SE (European patent), US.

Published
With international search report.

(54) Title: ELEMENT MOVING DEVICE

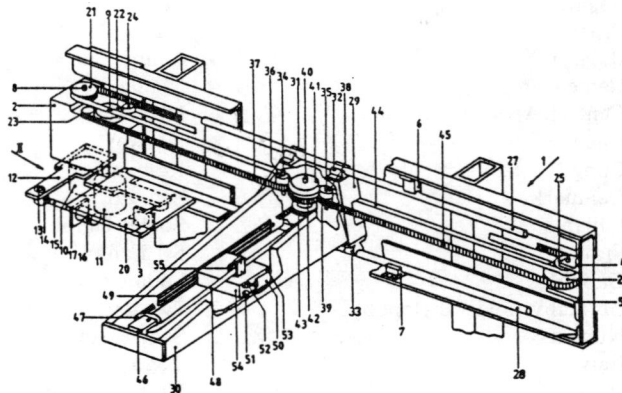

(57) Abstract

The element moving device includes a supporting structure (29, 30) movable in the X-direction on guide rods (27, 28) and carrying an axle (40) on which an assembly comprising three gear wheels (41, 42, 43) is freely rotatable. Two wheels (41, 42) are engaged by corresponding conveyor belts (44, 45) extending in the X-direction and able to control the rotation or standstill of the axle. The third wheel (43) is coupled to a carriage (54) movable in the Y-direction by a third conveyor belt (48) and carrying the element to be displaced. Each of the two conveyors (41, 42) is controlled by a pair of stepper motors (10, 16) which allow small and accurate displacements to be realized.

PCT

世界知的所有権機関
国際事務局

特許協力条約に基づいて公開された国際出願

(51) 国際特許分類 ⁴ H03F 1/34, 1/08, 3/189	**A1**	(11) 国際公開番号	**WO 86/ 02214**
		(43) 国際公開日	1986年4月10日 (10.04.86)

(21) 国際出願番号	PCT/JP85/00541	(74) 代理人
(22) 国際出願日	1985年9月30日 (30.09.85)	弁理士 中尾敏男, 外 (NAKAO, Toshio et al.)
(31) 優先権主張番号	特願昭59-205816	〒571 大阪府門真市大字門真1006番地
	特願昭60-32002	松下電器産業株式会社内 Osaka,(JP)
(32) 優先日	1984年10月1日 (01.10.84)	(81) 指定国
	1985年2月20日 (20.02.85)	DE,GB,US.
(33) 優先権主張国	JP	添付公開書類
		国際調査報告書

(71) 出願人 (米国を除くすべての指定国について)
松下電器産業株式会社
(MATSUSHITA ELECTRIC INDUSTRIAL CO., LTD.) [JP/JP]
〒571 大阪府門真市大字門真1006番地 Osaka,(JP)

(72) 発明者 ; および
(75) 発明者/出願人 (米国についてのみ)
臼井 晶 (USUI, Akira) [JP/JP]
〒569 大阪府高槻市津之江町3-8-24 Osaka,(JP)
山田 忠 (YAMADA, Tadashi) [JP/JP]
〒618 大阪府三島郡島本町広瀬4-22-27 Osaka,(JP)
久保一彦 (KUBO, Kazuhiko) [JP/JP]
〒618 大阪府三島郡島本町若山台1-5-8-102 Osaka,(JP)
永井裕之 (NAGAI, Hiroyuki) [JP/JP]
〒567 大阪府茨木市山手台6-8-24 Osaka,(JP)

(54) Title: HIGH-FREQUENCY AMPLIFIER
(54) 発明の名称 高周波増幅装置

(57) Abstract

A high-frequency amplifier used for amplifying high-frequency signals in a TV tuner circuit or a satellite broadcast receiver. Emitter-grounded amplifiers (Q_1, Q_2) are d-c coupled in at least two stages to constitute an integrated circuit. An emitter resistor (R_{13}) is connected between ground and the emitter of the emitter-grounded amplifier (Q_2) of the latter stage, the emitter of the emitter-grounded amplifier (Q_2) of the latter stage is connected through a bonding wire to an externally connecting terminal of the integrated circuit, and a filter circuit (11) is provided between this terminal and the external ground, the filter circuit (11) containing an inductance of the bonding wire so as to exhibit a small impedance for particular frequencies.

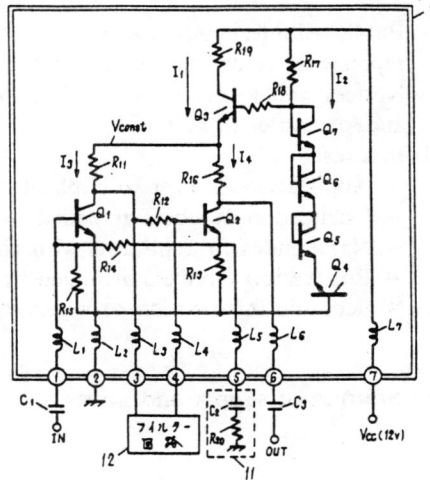

12 -- filter circuit

II Patent Cooperation Treaty documents

The International Bureau publishes the specification of each PCT application eighteen months after its priority date, or earlier if requested by the applicant. The specification, known as a pamphlet, has a standardized front page and also includes the search report if, as is usual, it has been established by the date of publication.

The numbers of the PCT applications are accorded by the receiving office and take the form PCT/US85/12345. The letters PCT are followed by the two letter country code of the receiving office (based on WIPO standard ST3) and the year of filing, and the five digits indicate the number of the application received by the office that year.

A new sequence of publication numbers is started each year prefixed by the year of publication, e.g. 85/00123. The first specification was published in 1978.

III Official gazette

PCT Gazette   1978 –   every two weeks (subject to change)

There are two editions, in English and French respectively.

The gazette has four sections: –

I   Published international applications: bibliographic details, an abstract and representative drawing of all pamphlets published concurrently.
II Notices and information relating to published international applications and/or entries in section I.
III Indexes.
   – Application number to publication number.
   – Publication numbers accorded to designated states.
   – Name index to applicants with corresponding publication number.
   – Publication numbers arranged by IPC.
IV Notices and information of a general character.

IV Sources of supply and prices

	Prices from 1987 SFr
PCT Gazette, either language	
annual subscription, surface mail	440
annual subscription, airmail	515
single issue, surface mail	18
single issue, airmail (except to the Americas or Japan)	20

PCT pamphlets (specifications)
single copy, surface mail   11
single copy, airmail   13

Obtainable from
WIPO
PO Box 18
CH-1211 Geneva 20
Switzerland

Orders and enquiries relating thereto from residents of the Americas and all dependencies of the United States should be addressed to the following sales agent of WIPO: −

Pergamon InfoLine Inc.
1340 Old Chain Bridge Road
McLean, Virginia 22101
USA
Tel: (703) 442-0900
Telex: 90-1811

Orders and enquiries relating thereto from residents of Japan should be addressed to the following sales agent of WIPO: −

Hatsumei Kyokai (JIII)
9-14 Toranomon
2-Chome, Minato-ku
Tokyo 105
Japan
Tel: (03) 502-0511
Telex: 0222-2947 IVENA J

PATENT OFFICE

TOKKYOCHO
4-3 Kasumigaseki 3-chome
Chiyoda-ku
Tokyo 100

I   Patent documents

Patent specifications are published unexamined eighteen months after their
priority date and examination is on request up to seven years after the filing
date. After examination the specification is published a second time for opposition. The procedure, including examination, is similar for utility models.

The specifications are not published singly but bound into volumes (gazettes) of
convenient size each having its own title page and contents list.

The titles, in Japanese and transliterated, are: –

A 公開特許公報　　　　　　　Patent specifications published
　Kôkai tokkyo kôhô　　　　before examination (A)

B 特許公報　　　　　　　　　Patent specifications published
　Tokkyo kôhô　　　　　　　after examination (B)

U 公開実用新案公報　　　　　Utility model specifications
　Kôkai jitsuyô shin'an kôhô　published before examination (U)

Y 実用新案公報　　　　　　　Utility model specifications
　Jitsuyô shin'an kôhô　　　published after examination (Y)

S 意匠公報　　　　　　　　　Registered design applications
　Ishô kôhô　　　　　　　　published before examination (S)

The cover page of *Kôkai tokkyo kôhô* is illustrated and the other gazettes have
a similar layout. The detail on the page is as follows: –

A Publication date, the first two figures are the year of the relevant Japanese
　era, currently *Shôwa*, which started in 1926, so year 61 corresponds to 1986
　in the Gregorian calendar.

A   昭和61（1986）. 2. 7発行

B   # 公開特許公報

C   61（1986）— 283 〔24033〕

D   第 7 部 門
第3区分（電子、通信関係）

E   7 (3) — 23〔964〕

F   特許出願公開　昭61— 28201〜28300

日 本 国 特 許 庁

⑲ 日本国特許庁（JP）　⑪ 特許出願公開

⑫ 公開特許公報（A）　昭61－28243

a　b

⑸Int.Cl.⁴　識別記号　庁内整理番号　㊸公開　昭和61年(1986) 2月7日

H 04 B 9/00　S－6538－5K

H 03 G 3/30　B－7210－5J

c　d　e

審査請求 有　発明の数 1（全4頁）

㊹発明の名称　光受信装置

㉑特　願　昭60－91296

㉒出　願　昭60(1985) 4月30日

優先権主張　㉜1984年5月4日㉝西ドイツ(DE)㉛P3416493.6

⑫発 明 者　ハンス・マルチン・ギユントナー　ドイツ連邦共和国，7141 ベンニンゲン，リービツヒシユトラーセ 4／1

㉑出 願 人　スタンダード・エレクトリツク・ローレンツ・アクチエンゲゼルシヤフト　ドイツ連邦共和国，7000シユツツトガルト－40，ローレンツシユトラーセ 10

⑭代 理 人　弁理士 鈴江 武彦　外2名

明　細　書

1．発明の名称

光受信装置

2．特許請求の範囲

（1）光減衰装置と、光－電気変換装置と、増幅装置と、自動レベル制御回路とを具備している光受信装置において、

光減衰装置によつて導入される減衰が電気的に変化可能に構成されていることを特徴とする光受信装置。

（2）光減衰装置（1）が自動制御システムの制御素子であり、その他の素子が光－電気変換装置（2）と、増幅器装置（3）と、自動レベル制御回路（4）とであることを特徴とする特許請求の範囲第1項記載の光受信装置。

（3）光減衰装置（1）の他に光－電気変換装置（2）と増幅器装置（3）の少なくとも一方が自動制御システムの制御素子であり、操作された変数が自動レベル制御回路（4）において生成されることを特徴とする特許請求の範囲第2項記載の光受信装置。

（4）光減衰装置（1）が半導体または$LiNbO_3$を基体とする電気光変調装置であることを特徴とする特許請求の範囲第1項乃至第3項のいずれか1項記載の光受信装置。

（5）光減衰装置（1）が光学的コネクタ中に設けられていることを特徴とする特許請求の範囲第1項乃至第4項のいずれか1項記載の光受信装置。

（6）光減衰装置（1）、光－電気変換装置（2）、増幅器装置（3）および自動レベル制御回路（4）が単一のチップ上に集積されていることを特徴とする特許請求の範囲第1項乃至第5項のいずれか1項記載の光受信装置。

3．発明の詳細な説明

［発明の技術分野］

この発明は、光減衰装置と、光－電気変換装置と、増幅装置と、自動レベル制御回路とを具備している光受信装置に関するものである。

［発明の技術的背景］

通常の回路においては光－電気変換装置に後続

B Title of the gazette i.e. *Kôkai tokkyo kôhô*.
C Year of publication, 61 (1986), issue number in that year, 283, and cumulative issue number since the beginning of the series, 24033.
D Title of the sub-series, in this case, electronics, communications, and number of the subject group 7 (3).
E Issue number in the sub-series, 23, and cumulative issue number since the beginning of the sub-series, 964.
F Serial numbers of specifications in the issue.

The presentation and layout of Japanese specifications are now similar to those in European languages. The INID code is applied and from 1980 only the IPC is used. The claims are at the beginning of the text and the drawings at the end. In the case of unexamined utility model applications only the claims, an explanation of the drawings and the drawings are published. The full text of the specification is published after examination.

The first page of the specification of an unexamined patent application is shown. Heading data not identified by the INID code are: –

a additional search terms
b office references
c indication of request for examination
d number of claims
e number of pages

The serial numbering system of Japanese specifications is confusing as all four series begin each year at number 1 preceded by the year of registration, Japanese style, e.g. 61-1234, so that there may exist four documents bearing the same number. It is therefore essential when ordering copies or citing references that the type of publication be quoted, e.g. unexamined patent specification or *Kôkai tokkyo kôhô*, as well as the serial number.

Furthermore, the national application number at INID code (21) is presented in the same form i.e. year followed by serial number. A concordance table application no./publication no. is published by JAPIO.

After grant patents and utility models are allocated a further number in separate *continuous* sequential seven figure series.

II Official gazette

Patent office gazette (Tokkyochô kôhô)

This is published in a number of separate parts with different coloured covers. The principal sections are: –

Patents granted with serial numbers – white cover
Utility models granted with serial numbers – pink cover

Trade marks stages of progress – yellow cover
Patent applications on which examination has been requested – grey cover
Utility model applications on which examination has been requested –
mauve cover
Annual report – white cover

III Patent abstracts of Japan March 1977 –

This publication consists of abstracts in English, with full heading data, of
published unexamined Japanese patent applications *(Kôkai tokkyo kôhô)*. The
selection of specifications is made in "technical fields where a high incidence of
patent applications between Japan and foreign countries usually exists". In
practice about half the total number of published specifications are abstracted
but none from non-residents which will have been abstracted elsewhere.

Each issue contains 500 abstracts in one of the following subject groups: –

General and mechanical	M	Field
Chemical	C	Field
Physical	P	Field
Electrical	E	Field

IV Sources of supply

Specifications, gazettes and *Patent abstracts of Japan* may be obtained from the
following two organizations: –

JAPIO, Japan Patent
Information Organization
Bansui Bldg
5-16 Toranomon 1-chome
Minato-ku
Tokyo, 105

Hatsumei Kyokai
Japan Institute of Invention
and Innovation
9-14 Toranomon 2-chome
Minato-ku
Tokyo, 105

JAPIO, Japan Patent Information Organization　　　　　　**JP**
Bansui Bldg
5-16 Toranomon 1-chome
Minato-ku
Tokyo, 105

JAPIO was established as a non-profit making organization in August 1985 by rearranging the work of JAPATIC (Japan Patent Information Center) and the patent information services of JIII (Japan Institute of Invention and Innovation) so as to meet the increasing demand for information about patents.

Its objective is to collect, store and process national and foreign patent information and it maintains a library of official patent publications of many countries and also provides translations and indexes.

A database, PATOLIS (Patent Online Information System), has been established containing bibliographic data relating to Japanese patents, utility models, designs and trade marks, currently from about 1965, though some earlier patent and trade mark data are accessible. Abstracts in English of Japanese unexamined specifications have been available online since 1985.

SOVIET UNION   SU

PATENT OFFICE

USSR STATE COMMITTEE FOR INVENTIONS AND DISCOVERIES
(Gosudarstvenny komitet SSSR po delam izobreteny i otkryty)
Malyi Tcherkasski Pereulok 2/6
Moscow
GSP, 103621

I   General information

The State Committee for Inventions and Discoveries has much wider respon-
sibilities than patent offices in Western Europe as it is concerned with all aspects
of inventive activity from the registration of inventions to their national and
foreign exploitation. The Statute, last amended in 1978, covers not only patents
and inventors' certificates but also "discoveries", i.e. previously unknown
qualities of the material world which increase scientific knowledge, and 'rationalization proposals" for the improvement of products and processes.

The dissemination of atent information is the work of subsidiary institutes. The
Research Institute of Patent Information prepares evaluation reports on both
national and foreign patent literature. The Productive Polygraphic Enterprise
Patent assesses applications before passing them on to the examining section and
also provides services to industry.

The All-Union Patent and Technical Library (*Vsesoiuznaia patentnotekhnicheskaia biblioteka*), Berezhkovskaya nab 24, Moscow 121857, has all national and
foreign patent specifications and many books and periodicals.

II   Patent documents

Two forms of protection are provided viz. inventors' certificates through which
inventions are assigned to the State which assumes responsibility for their exploitation and conventional patents. In practice nearly all national applications
are for inventors' certificates as most inventors are employed by the State and
most conventional patents are owned by foreigners. Inventors' certificates do,
however, provide protection for a much wider range of inventions such as new
varieties of plants and breeds of farm animals and also for methods of preven-
tion, diagnosis and treatment of human diseases.

186

СОЮЗ СОВЕТСКИХ
СОЦИАЛИСТИЧЕСКИХ
РЕСПУБЛИК

(19) SU (11) 1261572 A1

(51) 4 A 01 B 79/00

ГОСУДАРСТВЕННЫЙ КОМИТЕТ СССР
ПО ДЕЛАМ ИЗОБРЕТЕНИЙ И ОТКРЫТИЙ

ОПИСАНИЕ ИЗОБРЕТЕНИЯ

К АВТОРСКОМУ СВИДЕТЕЛЬСТВУ

(21) 3687515/30-15
(22) 05.12.83
(46) 07.10.86. Бюл. № 37
(71) Украинская ордена Трудового Красного
Знамени сельскохозяйственная академия
(72) М. А. Зеленский, О. Б. Зеленская-
Новоминская и В. М. Винничук
(53) 631.514(088.8)
(56) Авторское свидетельство СССР
№ 132445, кл. A 01 B 79/00, 1958.

тивного побегообразования. Первый раз рас-
тения окучивают после появления у них 2—3
стеблей. Повторные окучивания осуществля-
ют каждый раз после появления из пазуш-
ной почки каждого из уже образованных
стеблей новых 1—2 стеблей. Второй раз окучивают после появления 6—7 стеблей. Третий
раз — после появления 9—12 стеблей путем
присыпания основания стеблей слоем почвы.
В трех примерах окучивание всходов ози-
мой пшеницы в осенний период вегетации

СОЮЗ СОВЕТСКИХ
СОЦИАЛИСТИЧЕСКИХ
РЕСПУБЛИК

(19) SU (11) 1261571 A3

(51) 4 G 05 F 1/14, H 02 M 5/12

ГОСУДАРСТВЕННЫЙ КОМИТЕТ СССР
ПО ДЕЛАМ ИЗОБРЕТЕНИЙ И ОТКРЫТИЙ

ОПИСАНИЕ ИЗОБРЕТЕНИЯ

К ПАТЕНТУ

(21) 3564996/24-07
(22) 03.03.83
(31) 647/82
(32) 03.03.82
(33) HU
(46) 30.09.86. Бюл. № 36
(71) Электроакустикай Дьяр (HU)
(72) Томаш Фачади и Лайош Торньи
(HU)
(53) 621.314.2(08)
(56) Патент США №
кл. G 05 F 1/14,
    Патент США № 4(
кл. G 05 F 1/14, 1

Inventor's certificate A1
Patent A3

Цель изобретения — повышение точнос-
ти поддержания выходного напряжения
и снижение стоимости. Для стабилиза-
ции сетевого напряжения изменяют
коэффициент трансформации регулиру-
ющего трансформатора 10 изменением
состояния электронной системы 9 ком-
мутации. Электронная система 9 комму-
тации управляется сигналом, пропорци-
ряжению с выхода
аторов. Сигнал,
ых компараторов
перемещению
в действие

ОФИЦИАЛЬНЫЙ
БЮЛЛЕТЕНЬ
ГОСУДАРСТВЕННОГО
КОМИТЕТА СССР
ПО ДЕЛАМ
ИЗОБРЕТЕНИЙ
И ОТКРЫТИЙ

ОТКРЫТИЯ ИЗОБРЕТЕНИЯ

ИЗДАЕТСЯ С 1924 ГОДА
ВЫХОДИТ ЧЕТЫРЕ РАЗА В МЕСЯЦ

СВЕДЕНИЯ, ПОМЕЩЕННЫЕ В НАСТОЯЩЕМ БЮЛЛЕТЕНЕ,
СЧИТАЮТСЯ ОПУБЛИКОВАННЫМИ 23 МАРТА 1987 г.

ВСЕСОЮЗНЫЙ НАУЧНО-ИССЛЕДОВАТЕЛЬСКИЙ
ИНСТИТУТ ПАТЕНТНОЙ ИНФОРМАЦИИ

МОСКВА·1987

11

Official gazette – Otkrytiia, isobreteniia

The examination, acceptance and publication procedure is similar for both inventors' certificates and patents. On the front page of the specifications are bibliographic data with both the Universal Decimal Classification (UDC) and the IPC, claims, a list of prior art documents and a drawing. The specifications for inventors' certificates are relatively brief.

III Official gazette

Discoveries, inventions (Otkrytiia, isobreteniia)   1924 –

48 issues a year. Since 1983 designs and trade marks are in a separate volume.

Contents

– Abstracts or claims of granted inventors' certificates
– Abstracts or claims of granted patents
– Abstracts or claims of inventors' certificates not previously published
– Classified list of specification numbers
– Concordance, application no./publication no.
– Occasional lists of discoveries and rationalization proposals

IV Sources of supply

Official publications available from

All-Union Patent and Technical Library
(Vsesoiuznaia patentnotekhnicheskaia biblioteka)
Berezhkovskaya nab 24
Moscow, 121857

PATENT OFFICE

UNITED STATES PATENT AND TRADEMARK OFFICE (USPTO)
Commissioner of Patents and Trademarks
Washington DC 20231
USA

I   General information

The USPTO is an agency of the United States Department of Commerce which administers the patent and trademark laws.

Total staff: 3200
Patent professionals: 1250

II   Patent documents

Patent specifications, with a very few exceptions, are published after grant and there is no early publication or advance information. There are no utility models but the conventional patent is sometimes referred to as a "utility patent" to distinguish it from a plant patent or design patent.

On the front page of the specification are bibliographical data with both IPC and the US classification, a list of references cited which may include journal and book references as well as patents, and also an abstract with a related significant drawing.

US patents are often cited by their application number and there is a concordance application no./publication no. from 1968 published by JAPATIC (now JAPIO, Japan Patent Information Organization).

Other patent publications are: –

Reissue patents   A new patent in a separate series prefixed Re may be issued when the original is in error or new prior art becomes known and claims are extended.

Reexamination certificates   These may be issued following new examination which may be requested by any interested party.

United States Patent [19]

Lee et al.

[11] Patent Number: 4,584,504

[45] Date of Patent: Apr. 22, 1986

[54] **INTEGRATED CIRCUIT FOR DRIVING A D.C. MOTOR HAVING OPERATIONAL MODES**

[75] Inventors: Bang W. Lee, SuWon; Sung I. Hong, YangPyonKun, both of Rep. of Korea

[73] Assignee: Samsung Semiconductor and Telecommunications Co., Ltd., KyoungSangBukDo, Rep. of Korea

[21] Appl. No.: 649,280

[22] Filed: Sep. 11, 1984

[30] **Foreign Application Priority Data**

May 10, 1984 [KR] Rep. of Korea .... 84-2520[U]

[51] Int. Cl.⁴ H04Q 7/02

[52] U.S. Cl. 318/16; 318/293; 340/825.69

[58] Field of Search 318/16, 293; 340/825.69, 825.72, 870.28; 307/270, 262, 255, 254

[56] **References Cited**

U.S. PATENT DOCUMENTS

4,275,394	6/1981	Mabuchi et al.	340/825.69 X
4,349,986	9/1982	Tsukuda	340/825.69 X
4,488,094	12/1984	Min et al.	318/16
4,490,655	12/1984	Feldman	307/254 X

OTHER PUBLICATIONS

Giles et al, "Two-Chip Radio Link Pilots Toys and Models", Electronics, Jun. 5, 1980, vol. 53, No. 13, pp. 145-149.

Primary Examiner—William M. Shoop, Jr.
Assistant Examiner—Bentsu Ro
Attorney, Agent, or Firm—Bacon & Thomas

[57] **ABSTRACT**

The present invention relates to an integrated circuit for driving a d.c. motor with radio control comprising a receiving circuit for receiving and detecting certain signals transmitted from a transmitter, an amplifier for amplifying an output signal of said receiving circuit, a peak detector for converting the said amplified audio signal into a d.c. voltage, a comparator which have a hysteresis character dependent on the output level of the peak detector, a voltage regulating circuit supplying a stabilized voltage into all other components, and a direction control circuit to generate logic control signals deciding actual operation mode of the d.c. motor and motor driving circuits to produce motor driving signals by the output signal of the direction control circuit.

6 Claims, 21 Drawing Figures

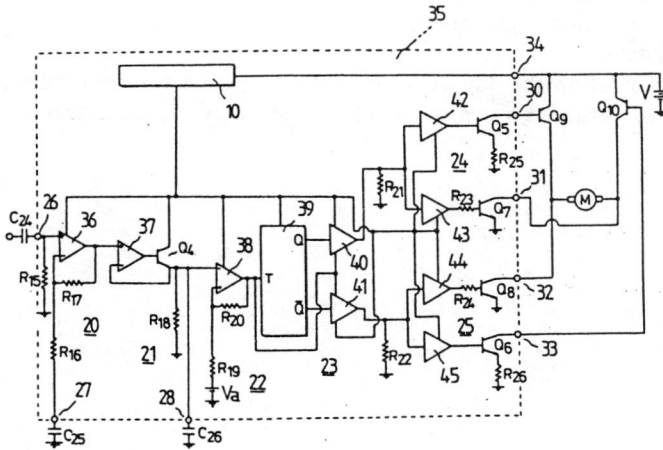

Defensive publication   The applicant waives patent rights and there is no examination. Abstracts and drawings are published.

Statutory invention registration   Introduced in 1985 the specification is published in a separate sequential series prefixed H without examination and the inventor waives the right to receive a patent within a certain period.

Plant patents   This new series was started in 1931 for new varieties of plant. The specification comprises heading data, one or more glossy photographs and a detailed description of the plant.

III Official gazette

Official gazette of the United States Patent and Trademark Office, 1872 – ,   weekly.

Contents

Office notices, new legislation, etc
List of government inventions available for licensing
Heading data and the first or most significant claim and drawing for: –

- Reexamination certificates
- Defensive publications
- Statutory invention registrations
- Reissue patents
- Patents granted on the day of issue of the *Gazette* arranged numerically in three broad subject groups, General and Mechanical, Chemical and Electrical.
- Designs

IV Sources of supply and prices

Official gazette

Superintendant of Documents
US Government Printing Office
Washington DC 20402

Prices from summer 1984

	Prices $
Annual subscription, USA, 1st class mailing	375.00
Annual subscription, foreign, 4th class mailing	337.00
Single copies, USA	13.00
Single copies, foreign	16.25

Specifications

Commissioner of Patents and Trademarks
Washington DC 20231

Prices from November 1985

Patent specifications, per copy	1.50
Plant patent specifications, per copy	6.00

V Public services

The USPTO Library has readily available all official gazettes and all post 1920 foreign patent documents needed to meet the minimum documentation requirements of the Patent Cooperation Treaty either as printed copy or in microform. Other foreign patent documents are either in the Library or in store.

CHEMICAL ABSTRACTS SERVICE (CAS)

2540 Olentangy River Road
PO Box 3012
Columbus, Ohio 43210
USA

General information

The journal *Chemical Abstracts* began in 1907 and covers a broad field of chemistry and chemical engineering. It is published weekly in both printed copy and in microform and there is also the option of acquiring printed volume indexes and weekly issues in microform. A complete backfile 1907 – 85 is available in microform.

More than 450,000 abstracts are now published annually including nearly 100,000 patent specifications of which currently about half are of Japanese origin.

Coverage

All chemical patents from Australia, Austria, Belgium, Brazil, Canada, France, German Democratic Republic, Germany Federal Republic, India, Israel, Japan, Netherlands, Romania, South Africa, Soviet Union, Switzerland, United Kingdom and United States of America.

Chemical patent specifications from the European Patent office and those published under the Patent Cooperation Treaty.

Chemical patents issued to individuals and organizations resident in Czechoslovakia, Denmark, Finland, Hungary, Norway, Poland, Spain and Sweden.

The patent index includes entries for all newly abstracted patent documents on an invention, cross references to the first abstracted document on an invention and a listing of all patent documents related to a particular invention, i.e. patent families.

CAS online

The CA file, available on STN International, contains all CA references since 1967 with complete abstracts available from mid – 1975. The service has now been extended with the added facility of searching abstract text. A comprehensive range of user aid manuals is available.

Training, support and assistance to users of CAS is provided by the following centres in Europe: –

For Denmark, Finland, Ireland, Netherlands, Norway, Sweden and the United Kingdom: –
> The Royal Society of Chemistry
> The University
> *Nottingham NG7 2RD*
> Tel: 0602-507411
> Telex 37488

For France, Monaco and many African nations: –
> Centre National de l'Information Chimique
> La Maison de Chimie
> 28 ter rue Saint Dominique
> *75007 Paris*
> Tel: (1)45-51-37-40
> Telex: 202634

For CAS online in
> Austria, Federal Republic of Germany and Switzerland: –
> Fachinformationszentrum Chemie Gmbh
> Postfach 126050
> Steinplatz 2
> *1000 Berlin 12*
> Tel.: 030/31 90 03-0
> Telex: 181255
> Federal Republic of Germany

For CAS printed and microform services in Austria, Federal Republic of Germany and Switzerland: –
> VCH Verlagsgesellschaft mbH
> CA Vertrieb
> Postfach 1260/1280
> *D-6940 Weinheim*
> Tel: 06201/602-231
> Telex: 465516
> Federal Republic of Germany

The STN International Help desk can be contacted at: –
> STN Karlsruhe
> c/o FIZ Energie, Physik, Mathematik GmbH
> Postfach 2465
> *D-7500 Karlsruhe 1*
> Tel.: 7247/82-4566
> Telex: 7826487

Computer readable files are available at the following organizations: –

Belgium   Bibliothèque royale Albert 1er
Boulevard de l'Empereur 4
B-1000 Bruxelles
Tel. (02)51 36 180
Telex (2) 21157

France   Télésystèmes/Questel
83-85 boulevard Vincent Auriol
75013 Paris
Tel. 45 82 64 64
Telex 204594

Italy   European Space Agency
Via Galileo Galilei
00044 Frascati
Tel. 39(6) 940 11
Telex 610637

Spain   Instituto de Información y Documentación
en Ciencia y Tecnologia
Joaquin Costa 22
Madrid 6
Tel. 26 14 808
Telex 22628

Sweden   Kungliga Tekniska Högskolans
(Royal Institute of Technology)
Bibliotek
Valhallavägen 81
S 10044 Stockholm 70
Tel. (08)78 70 000
Telex 10389

Switzerland   Data-Star
12 Quai de la Poste
CH 1204 Geneva
Tel. 031-65 95 00

DERWENT PUBLICATIONS LTD.

Rochdale House
128 Theobalds Road
London WC1X 8RP
Tel: 01-242 5823
Telex: 267487 DERPUB
Telecopier: 01-405 3630
Cable: DERWENTINF LONDON

I General information

Derwent analyses, classifies, indexes, abstracts and codes patent documents from each of the world's principal issuing authorities. The information which results is provided in the form of easy-to-use packaged services. These include printed materials in order of country and subject matter, microform, and as an online database – 'World Patents Index'.

Derwent acquires the patent specifications and official gazettes from 29 patent issuing authorities immediately they are published, making special arrangements for fast delivery to London.

II Coverage

Australia (1963 – 69, 1983 –)	Japan	Additional sources:
Austria (1975)	Luxembourg (1984)	
Belgium	Netherlands	*Research*
Brazil (1976)	Norway (1974)	*Disclosure (1978)*
Canada	Portugal (1974)	
Czechoslovakia (1975)	Romania	*International*
Denmark (1974)	South Africa	*Technology*
Finland (1974)	Spain (1983)	*Disclosures (1984)*
France	Sweden (1974)	
Germany (BRD)	Switzerland	
Germany (DDR)	USA	
Great Britain	USSR	
Hungary (1975)	EPO (Europe) (1978)	
Israel (1975)	PCT (World) (1978)	
Italy (1978)		

Dates in parenthesis indicate year of first inclusion when different from commencement of a specific service.

13,000 patent documents are processed each week, beginning with input of their bibliographic details to Derwent's computer masterfile.

Subject classification of the incoming documents enables them to be sent to technical specialists possessing the appropriate combination of subject knowledge and language expertise. Following clearly defined guidelines, an informative English language title is produced to replace the brief official one and an abstract is written which incorporates all the significant features and main claims of the original document.

Derwent classifies subject matter into sections, each identified alphabetically.

III Services and publications

CHEMICAL PATENTS INDEX (CPI)

This service commenced in 1970 as Central Patents Index. The name was changed to Chemical Patents Index in 1986. It incorporates Farmdoc (CPI Section B) for 1963, Plasdoc (CPI Section A) from 1966 and Agdoc (CPI Section C) from 1975.

CPI provides a full range of weekly abstracts publications for current awareness, extensive indexing including chemical structure/substructure retrieval, plus numerous ancillary products and features.

About 6,000 patents are covered by Derwent weekly in CPI.

Alerting Abstracts Bulletins

For each of the twelve Sections of CPI (A-M), Alerting Abstracts Bulletins are available in either Country Order or Classified Order. The Country versions contain abstracts in alphabetical order of country and in patent number sequence for both basics and equivalents.

The 'Classified' versions appear one week later than the 'Country' editions. The 135 CPI Classes provide helpful subject headings to aid scanning.

Both of them contain Patentee, Accession Number and Patent Number indexes. The 'Country' version also has a CPI Class Index.

Documentation Abstracts Journals

Unique to the CPI service is a second type of abstract — the Documentation Abstract.

These contain much extra detail and they are published once only for each invention, in accession number sequence. They are designed to be the prime source of reference for retrospective searching, available from the commencement of each Section.

Twelve Documentation Abstracts Journals (for Sections A-M) are published weekly, each with a sectional Accession Number Index, Patent Number Index, Patentee Index and Manual Code Index.

ELECTRICAL PATENTS INDEX (EPI)

EPI was started in 1980 and is designed to meet the special requirements of the electrical and electronics industries.

However, electrical and electronics patents have been covered and are searchable from 1974.

For convenience of use, EPI is divided into six subject Sections (S-X).

About 4,300 patents are covered weekly from all 29 patent issuing authorities in the Derwent system.

Alerting Abstracts Bulletins

For each of the six Sections, two types of weekly Alerting Abstracts Bulletins containing abstracts are available: a Country Order and a Classified Order version.

The Alerting Abstracts Bulletin (Classified) apears just one week later than the 'Country' version and contains essentially the same information, but organised by EPI Class within each of the six editions.

Both versions contain sectional Patentee, Accession Number and Patent Number Indexes. In addition, the 'Country' version has a Class Index, and the 'Classified' version a Manual Code Index.

To cope with the very large numbers of Japanese electrical and electronics patents encountered, coverage is restricted to unexamined (Kokai) documents in Section H (Electricity) of the International Patent Classification (IPC). For most of the Kokai no abstract is available, only the informative Derwent title, supplemented by drawings from the specification.

EPI Profile Booklets

Unique within the Derwent services is a range of monthly EPI Profile Booklets, one for each of the 50 EPI subject classes.

WORLD PATENTS ABSTRACTS

Within WPA, there are two distinct categories of publication. The first is a series of journals, each presenting the patents abstracts of one country. The second category, WPA Journals by Subject, covers general, mechanical and electrical technology.

WPA – Printed Products by Country

WPA publications for individual countries are available for eleven of the patent issuing authorities covered by Derwent.

For a given country, where weekly output is large, the abstracts are divided into separate editions based upon the categories Ch (Chemical), P (General), Q

(Mechanical) and El (Electrical). Conversely, where output is small, as in the case of France and South Africa, the abstracts for these countries are combined in a single edition.

Belgian Patents Abstracts	Complete Edition
British Patents Abstracts	Chemical, General/Mechanical and Electrical Numerical Unexamined and Granted
European Patents Abstracts	Chemical, General/Mechanical and Electrical Numerical Unexamined and Granted
French Patents Abstracts (Includes South Africa)	Chemical only
German Patents Abstracts (Examined)	Chemical, General/Mechanical/Electrical
German Patents Abstracts (Unexamined)	Chemical, General/Mechanical/Electrical
Japanese Patents Abstracts (Unexamined)	Chemical only
Netherlands Patents Abstracts	Chemical only
PCT Patents Abstracts	Chemical only
Soviet Patents Abstracts	Chemical, General/Mechanical and Electrical
United States Patents Abstracts	Chemical, General/Mechanical and Electrical

The sectionalised order with clear section headings is much preferred by most users, since scanning can be selective. However, for patent departments where all the country's patents must be scrutinised unselectively, alternative publications in strict patent number sequence are also produced for British and European Patents, in separate Unexamined and Granted editions. The Granted version contains the main claim, with drawings where appropriate in place of the previously published Derwent abstracts.

WPA – Journals by Subject

A series of seven weekly WPA Subject Journals gives multiple country coverage of non-chemical technology. Of the 29 patent issuing authorities, only Japan is omitted.

There are seven titles:

P1-P3:	Human Necessities	S-T:	Instrumentation,
P4-P8	Performing		Computing
	Operations	U-V:	Semiconductors,
Q1-Q4:	Transport,		Components
	Construction	W-X:	Communications,
Q5-Q7:	Mechanical		Power
	Engineering		

WPI GAZETTE SERVICE

The WPI Gazette Service is essentially a low cost, abbreviated patents information service. It covers all technology, but does not contain abstracts.

Four WPI Gazettes are published each week – General (P), Mechanical (Q), Electrical (R) and Chemical (Ch).

The absence of abstracts in the WPI Gazette Service means that the Gazettes can be compiled very rapidly. They contain the very first published information on a particular patent anywhere within the Derwent services. This can be as little as five weeks after the publication of the original documents by the Patent Offices, and one week ahead of any of Derwent's abstracts publications.

In each Gazette there are two major 'indexes'. The Patentee Index lists the individual headings in sequence of Derwent's Company Codes and the IPC Index contains the same entries in subject sequence. The user does not need to have previous experience of the IPC, because Derwent provides simply worded subheadings at frequent intervals.

ONLINE ACCESS

The Derwent 'World Patents Index' database is available online as two files, designated 'WPI' covering the period 1963 – 80 and 'WPIL' covering 1981 to date. They are available through four of the world's leading online service suppliers: DIALOG Information Services Inc., SDC Information Services, Télésystèmes-QUESTEL and SDC of Japan Ltd.

The files contain information on 7 million patent documents corresponding to 3 million patent families. Over 30 search categories enable the user to search precisely and in-depth.

INTERNATIONAL PATENT DOCUMENTATION CENTER (INPADOC)

Möllwaldplatz 4
A-1040 Vienna
Austria
Tel: (0222) 65 87 84
Telex: 1-35337

I General information

INPADOC was established in Vienna by the Republic of Austria following an agreement between the World Intellectual Property Organization and the Austrian Government signed on 2 May 1972.

Its objective was the provision of a databank containing details of patent publications and arrangements were made with nearly all of the patent offices of the world for the prompt supply of the relevant bibliographic data.

II Coverage

At present the INPADOC databank covers more than 98 % of all patent documents published yearly worldwide and the total number of documents is more than 13 million.

Details of the relevant countries and starting dates are as follows: –

Issuing authorities

Argentina (AR)	1973 –	Finland (FI)	1968 –
Australia (AU)	1973 –	France (FR)	1968 –
Austria (AT)	1969 –	German Dem. Rep. (DD)	1973 –
Belgium (BE)	1964 –	Germany, Fed. Rep. (DE)	1967 –
Brazil (BR)	1974 –	Greece (GR)	1977 –
Bulgaria (BG)	1973 –	Hong Kong (HK)	1976 –
Canada (CA)	1970 –	Hungary (HU)	1973 –
China (CN)	1985 –	India (IN)	1975 –
Cuba (CU)	1974 –	Ireland (IE)	1973 –
Cyprus (CY)	1975 –	Israel (IL)	1968 –
Czechoslovakia (CS)	1973 –	Italy (IT)	1973 –
Denmark (DK)	1968 –	Japan (JP)	1973 –
Egypt (EG)	1976 –	Kenya (KE)	1975 –
ESARIPO (AP)	1984 –	Korea, Rep. (KR)	1978 –
European Patent Office (EP)	1978 –	Luxembourg (LU)	1945 –

Malawi (MW)	1973 –	Soviet Union (SU)	1973 –
Mexico (MX)	1981 –	Spain (ES)	1968 –
Mongolia (MN)	1973 –	Sweden (SE)	1968 –
Netherlands (NL)	1964 –	Switzerland (CH)	1969 –
New Zealand (NZ)	1978 –	Turkey (TR)	1973 –
Norway (NO)	1968 –	United Kingdom (GB)	1969 –
Philippines (PH)	1975 –	United States of America (US)	1968 –
Poland (PL)	1973 –	Viet Nam (VN)	1985 –
Portugal (PT)	1976 –	Yugoslavia (YU)	1973 –
Romania (RO)	1973 –	WIPO-PCT (WO)	1978 –
Singapore (SG)	1983 –	Zambia (ZM)	1968 –
South Africa (ZA)	1971 –	Zimbabwe (ZW)	1980 –

Bibliographic data

The following bibliographic data are stored in the databank: –

- country of publication
- code indicating the type of document (unexamined patent, examined patent, utility model, etc.)
- application number
- filing date of application
- publication number
- publication date, or if such date is not available, the date of the relevant entry in the official gazette
- IPC, all available codes
- country of priority document
- application number of priority document
- filing date of priority document

If available the following will also be indicated: –

- name of applicant, owner of the patent or successor at law
- inventor's name
- title of patent in original language or translated into English, French or German

III Services offered

Magnetic tape produced weekly containing the bibliographic data of all patent documents received by INPADOC during the previous week plus details of all existing patent family members for each patent document (INPADOC Family Data Tape/IFD)

Microfiche services updated and cumulated quarterly. After five years a new cumulation period begins.

- Patent Family Service (PFS) Identifies patents based on a common priority claim. Data arranged by priority country, priority date and priority application number.
- INPADOC Numerical List (INL) Accessory to and only available with PFS; data arranged in order of priority application numbers.
- Patent Classification Service (PCS) Documents arranged by IPC and within each IPC symbol by country of publication and date.
- Patent Applicant Service (PAS) Documents arranged in alphabetical order of standardized name of applicant and within each name entry by order of IPC.
- Patent Applicant Service, sorted by priorities (PAP) Similar to PAS but documents arranged within each name by order or priority (country, date, number).
- Patent Inventor Service (PIS) Documents arranged in alphabetical order of name of inventor
- Numerical Data Base (NDB) Basic bibliographic data arranged in numerical order for each country or organization covered.

The INPADOC Patent Gazette (IPG) is also on microfiches but issued weekly and is arranged in the following four parts each corresponding to one of the quarterly services described above. Equivalent patents (i.e. family members) are shown on all four parts.

> Selected Numerical Service (SNS) updates NDB
> Selected Classification Service (SCS) updates PCS
> Selected Applicant Service (SAS) updates PAS
> Selected Inventor Service (SIS) updates PIS

Other services on microfiches includes the Patent Register Service (PRS) which records changes in legal status of applications or granted patents published by eight organizations and Concordances (CON) from application number to publication number for 32 publishing bodies. INPADOC can also supply concordances prepared by JAPIO for United States and Japanese documents.

The CAPRI service (Computerized Administration of Patent Documents Reclassified According to IPC) is a special microfiche/tape service indicating the specific IPC symbol assigned to patent documents, which includes at least PCT minimum documentation up to 1972, on reclassification.

Online services

INPADOC online services are updated weekly and supplied by the following online vendors: −

INPADOC: PFS (with the relevant abstract numbers of CAS, Derwent and China) and PRS.

Pergamon InfoLine: Complete bibliographic database searchable by all searching criteria.

INKA: Complete bibliographic database searchable by all searching criteria for patent specifications in the German language (i.e. AT, DE, CH) from 1978.

Furthermore, INPADOC is the European representative for the Japanese Patent Data Base PATOLIS and is able to provide comprehensive online searches the results of which are presented in a form suitable for ready use by European customers.

Individual requests relating to the information contained in the databank can be dealt with and copies of relevant documents supplied on paper or 16 mm microfilm (more than 35,000 reels are available).

JOURNALS AND DATABASES

 I Journals with patents coverage, in English, French and German.
 II Databases entirely devoted to patents.
III Selected international journals.

I   Journals with patents coverage.

Nearly all the English language titles listed are essentially abstract journals many of which are accessible online and the relevant database is cited. The scope of such databases is constantly being augmented and many include journals not listed here which do abstract a few significant patents in a narrow subject field.

Although most abstract journals listed their source publications very few stated their selection criteria for patent specifications so the percentages quoted and countries covered were identified by scanning recent issues.

All the French language journals listed are technical publications devoting a few pages in each issue to new patents in the related field. Nearly all are derived from the abstract in the French official gazette, *BOPI − Brevets d'invention*. One journal reproduces extracts from the entry in *Chemical Abstracts* while patents occupy about half the pages of *Verres et réfractaires*, (about 700 patents a year).

The three German language journals cited contain a high percentage of patents coverage in narrow subject fields.

The specialized journals of Derwent Publications and Wila-Verlag are dealt with in more detail in separate chapters as also is the Chemical Abstracts Service.

II   Databases entirely devoted to patents

The principal databases are listed with brief details of their coverage. More patent offices are making their databases accessible online to the public either direct or through one of the many online vendor organizations. The most recent countries to offer this facility are China (CHINAPATS), the Federal Republic of Germany (PATDPA), Japan (JAPIO) and Spain (CIBERPAT).

III Selected international journals

Only *Patent World* is added to this list and some journals include other aspects of industrial property.

I JOURNALS WITH PATENTS COVERAGE, IN ENGLISH, FRENCH AND GERMAN

Journals in English with patents coverage

(w)   weekly
(bw) bi-weekly, every two weeks
(sm) semi-monthly, twice a month

(m)   monthly
(bm) bi-monthly, every two months
(q)   quarterly, every three months

Title	Publisher	% patents	Country coverage	Database
Abstract Bulletin (m)	Institute of Paper Chemistry, Appleton, WI 54915, USA	40	Multi-national	PAPERCHEM
API Abstracts (w)	American Petroleum Institute, 156 William Street, New York 10038	50	Multi-national	APIPAT
Biotechnology Abstracts (sm)	Derwent Publications Ltd 128 Theobalds Road, London WC1X 8RP	60	Multi-national	BIOTECH
BioInvention (m)	OMEC International Washington, USA	100	US	
Cadmium Abstracts (q)	Cadmium Association, 34 Berkeley Square, London, W1X6AJ	10	EP, GB, US, WO	ZLC
Catalysts in Chemistry (m)	R H Chandler Ltd, Box 55, Braintree, Essex, CM7 6JT	80	Multi-national	
Chemical Abstracts (w)	Chemical Abstracts Service (CAS) PO Box 3012 Columbus Ohio 43210	20 (see separate entry)	Multi-national	STN Intl
Digest of Information and Patent Review (m)	British Glass Industry Research Association, Sheffield, S10 2UA	50	GB, US	
Food Science and Technology Abstracts (m)	Commonwealth Agricultural Bureaux, Farnham Royal, SL2 3BM, England	15	Multi-national	FSTA
Foods Adlibra (sm)	Foods Adlibra Publications, 9000 Plymouth Ave, Minneapolis, MN 55427, USA	5	US	FOODS ADLIBRA
Gas Abstracts (m)	Institute of Gas Technology, 3424 S State St, Chicago, IL 60616	10	US	

Title	Address		Coverage	
Industrial Diamond Review (bm)	De Beers Industrial Diamond Division, Ascot, SL5 9PX, England	25	Multi-national	
Information Science Abstracts (m)	Plenum Publishing Corp. 233 Spring St, New York, NY 10013	10	US	ISA
INIS Atomindex (sm)	International Atomic Energy Authority, Box 100, Vienna	10	Multi-national	INIS
Int J of Micrographics and Video Technology (q)	Pergamon Press Oxford		US, WO	
Journal of Synthetic Methods (m)	Derwent Publications Ltd, 128 Theobalds Road, London, W1X 6AJ	15	Multi-national	SDC
Lead Abstracts (q)	Lead Development Association, 34 Berkeley Square, London W1X 6AJ	15	EP, GB, US, WO	ZLC
McGraw Hill's Biotech Patent-watch (sm)	McGraw Hill, 1221 Avenue of the Americas, New York, NY 10020	100	EP, GB, WO	
Microbiology Abstracts Section A (m)	Cambridge Scientific Abstracts, 5161 River Road, Bethesda, MD 20816, USA	5	GB, US	IRL
Organometallic Compounds (bw)	R H Chandler Ltd, Box 55, Braintree, Essex, CM7 6JT	20	Multi-national	
Packaging Science and Technology Abstracts (bm)	International Food Information Service, 6000 Frankfurt/Main 71, Germany FR	20	Multi-national	PSTA

Title	Publisher	% patents	Country coverage	Database
Photographic Abstracts (bm)	Royal Photographic Society Milsom Street Bath BA1 1DN	45	EP, GB, US, WO	
Science and Technology Abstracts (m)	Food Research Association Leatherhead, Surrey, KT22 7RY, England	10	GB	FROSTI
Soils and Fertilizers (m)	Commonwealth Bureau of Soils, Farnham Royal, SL2 3BM, England	5	US	CAB
Tobacco Abstracts (bm)	Tobacco Literature Service, North Carolina State University, Raleigh, USA	10	US	
World Aluminium Abstracts (m)	Aluminium Association, 818 Connecticut Ave, Washington, DC	15	Multi-national	WAA
World Patents Abstracts (w)	Derwent Publications (see separate entry)	100	Multi-national	WPI, WPIL
World Surface Coatings Abstracts (sm)	Paint Research Association, Teddington, Middlesex, TW11 8LD	30	Mulit-national	WSCA
World Textile Abstracts (sm)	Shirley Institute, Manchester, M20 8RX	25	EP, GB, US	World Textiles
Zinc Abstracts (q)	Zinc Development Association, 34 Berkeley Square, London, W1W 6AJ	20	EP, GB, US, WO	ZLC

Journals in French with patents coverage

Biofutur (m)	29, rue Buffon, 75005 Paris
La France horlogère 11 issues a year	Société d'éditions Millot et Cie 20, rue Gambetta, 25000 Besancon
Galvano-Organo Traitements de Surface (m)	126, boulevard Péreire, 75017 Paris
L'industrie ceramique (m)	14, rue Falguière, 75015 Paris
L'industrie textile 11 issues a year	36, rue Ballu, 75009 Paris
Parfums, cosmétiques, arômes (bm)	5, rue Jules Lefèbvre, 75009 Paris
Revue francaise des corps gras (m)	Institut des corps gras, 10A, rue de la Paix, 75002 Paris
Revue technique des industries du cuir 10 issues a year	54, rue René Boulanger, 75010 Paris
Verres et réfractaires (bm)	Institut du verre, rue Michel Ange, 75016 Paris

Journals in German with patents coverage

Hochmolekularbericht (sm)	Bayer AG, Zentrale Forschung und Entwicklung Wiss. Information und Dokumentation, Leverkusen 1
Textilbericht (sm)	− same address as above −
Das Papier (m)	Papiermacher-Berufsgenossenschaft, Mainz 31

II DATABASES ENTIRELY DEVOTED TO PATENTS

Database	Subject	Coverage		Producer	Online vendor
CHINAPATS	All	CN Unex. patent abstracts	1985 –		InfoLine
CIBERPAT	All	ES	1968 –	RPI	RPI
CLAIMS					
PATENTS	All	US patent abstracts	1950 –	IFI Plenum Data	DIALOG
CITATION	All	Patents cited in US patents	1947 –	IFI Plenum Data	DIALOG
UNITERM	Chemical	US patents subject indexing	1950 –	IFI Plenum Data	DIALOG
COMPUTER-PAT	Digital data processing	US	1942 –	Pergamon	InfoLine
INPADOC		Multi-national			
	All	PFS, PRS	1968 –	INPADOC	INPADOC
	All	Full bibliographic and PFS	1968 –	INPADOC	InfoLine
PATENTE	All	AT, DE, CH	1978 –	INPADOC	INKA
INPANEW	All	Last 15 weeks and PFS		INPADOC	InfoLine
PATSDI	All	Last 6 weeks		INPADOC	INKA
INPI 1 (FPAT)	All	FR	1969 –	INPI	QUESTEL
INPI 2 (EPAT)	All	EP	1978 –	EPO/INPI	QUESTEL
INPI 3 (EDOC)	All	Multi-national, patent families	1969 –	INPI	QUESTEL
JAPIO	All	JP Unex. patent abstracts	1976 –	JAPIO	SDC
LEXPAT	All	US full text	1975 –	Mead Data	Mead Data
PATDATA	All	US abstracts	1971 –	BRS	BRS, IHS, Bertelsmann (for German language countries)
PATDPA	All	DE	1973 –	Deutsches Patentamt	STN
PATOLIS	All	JP	1955 –	JAPIO	INPADOC (in Europe)

Database	Subject	Coverage		Producer	Online vendor
PATOS	All	DE Offenlegungs-schriften	1968 –	WILA-Verlag Bertelsmann	Bertelsmann
PATSEARCH	All	US, WO abstracts	1970 –	Pergamon	InfoLine
SITADEX	Legal status	ES	1979 –	RPI	RPI
USCLASS	All	US	1790 –	Derwent	SDC
USPATENTS	All	US citations and claims	1970 –	Derwent	SDC
WPI/WPIL	All	Multi-national	1963 –	Derwent	DIALOG, QUESTEL, SDC

Bertelsmann Informations Service GmbH, Neumarkter Straße 18, D-8000
Munich 80, Federal Republic of Germany
BRS, 1200 Route 7, Latham, NY 12110, United States of America
Derwent Publications Ltd, Rochdale House, Theobalds Road, London
WC1X 8RP, United Kingdom
DIALOG Information Services Inc, PO Box 8, Abingdon, Oxfordshire
OX13 6EG, United Kingdom
IFI Plenum Data Corp. 302 Swann Avenue, Alexandria, VA 22301, United
States of America
IHS, Information Handling Services, 15 Inverness Way East, Englewood,
CO 80150, United States of America
InfoLine (Pergamon), 12 Vandy Street, London EC2A 2DE, United Kingdom
INKA, c/o Fachinformationszentrum, Energie, Physik, Mathematik, GmbH,
Eggenstein-Leopoldshafen 2, 7514 Karlsruhe, Federal Republic of
Germany
INPADOC, Möllwaldplatz 4, 1040 Vienna, Austria
INPI, Institut National de la Propriété Industrielle, 26 bis rue de Léningrad,
75800 Paris Cedex 08, France
JAPIO, Japan Patent Information Organization, Bansui Bldg, 1 – 5 – 16
Toranomon, Minato-ku, Tokyo, Japan
Mead Data Central International, International House, Suite 24, 1 St
Katharine's Way, London E1 9UN, United Kingdom
ORBIT Search Service, Bakers Court, 4th Floor, Baker Rd, Uxbridge,
Middlesex UB8 1RG, United Kingdom
Pergamon – see InfoLine
QUESTEL – see Télésystèmes
RPI, Registro de la Propiedad Industrial, Calle de Panama 1, 28036 Madrid,
Spain

SDC, System Development Corporation, – see ORBIT
STN International – PO Box 2465, D-7500 Karlsruhe 1, Federal Republic of
    Germany
Télésystèmes QUESTEL, 83 – 85 blvd Vincent Auriol, 75013 Paris, France
WILA-Verlag, Wilhelm Lampl KG, Landsberger Straße 191A, 8000 Munich 21,
    Federal Republic of Germany

III SELECTED INTERNATIONAL JOURNALS ON INDUST-RIAL PROPERTY

European Intellectual Property Review (EIPR)   1978 –   monthly
ESC Publishing, Oxford
Contains articles on the whole range of intellectual property not restricted to
Europe. Centre pages provide details of cases of special interest and new legisla-
tion under European Digest, International News and Overseas News.

*Gewerblicher Rechtsschutz und Urheberrecht. Internationaler Teil. (GRUR In-
ternational)*   1952 –   monthly, in German
Deutsche Vereinigung für Gewerblichen Rechtsschutz und Urheberrecht,
Weinheim
Articles and reports of cases in Germany, overseas and international, news of
events, conferences, changes in legislation, book reviews and bibliography of
recently published books and journal articles.

Industrial Property   1962 –   monthly
La propriété industrielle (in French)   1885 –
World Intellectual Property Organization, Geneva
Reports of WIPO meetings, news of international conventions, treaties, etc.,
and translations into English and French of industrial property legislation in
member countries.

International Review of Industrial Property and Copyright Law (IIC)
1969 –   bi-monthly (in English)
Max Planck Institute for Foreign and International Patent, Copyright and Competition Law, Munich
Presents "articles, decisions and other material of international importance in
the fields of Patents, Copyrights, Designs, Trademarks, Unfair Competition
and related Antitrust Problems". There are also books reviews and a section on
News and Information.

Les Nouvelles, Journal of the Licensing Executives Society
1966 –   quarterly
Licensing Executives Society (USA & Canada)
Contributed articles from authors world wide on licensing, the transfer of technology and industrial property rights.

Patent World   1987 –   monthly
Intellectual Property Publishing Ltd.
This "international journal for innovation professionals" has feature articles on many aspects of patenting worldwide from bioengineering in Japan to technology transfer, usually with cartoon type illustrations. There is also explanation and comment on new legislation, reports on symposia and book reviews, with a substantial section devoted to news of interesting patents, innovation, decisions and statistics.

Patents and Licensing   1971 –   bi-monthly (in English)
Japan Engineering News Inc. Tokyo
Mostly about Japanese patent office prodedures, court decisions, technology development and licensing. Not strictly international but has a regular topical article from "overseas correspondents".

World Patent Information   1979 –   quarterly
Pergamon
Described as "the international journal for patent documentation, classification and statistics of the Commission of the European Communities and the World Intellectual Property Organization" the content of each issue is provided by specialists from many countries. The journal is particulary concerned with the use of patent literature, exposing problems, describing new projects and the effects of new legislation.

INDEX